Robin Cousin

The Phantom Scientist

translated by Edward Gauvin

The MIT Press
Cambridge, Massachusetts
London, England

LIBRARY | BUILDING A | BUILDING B | BUILDING C

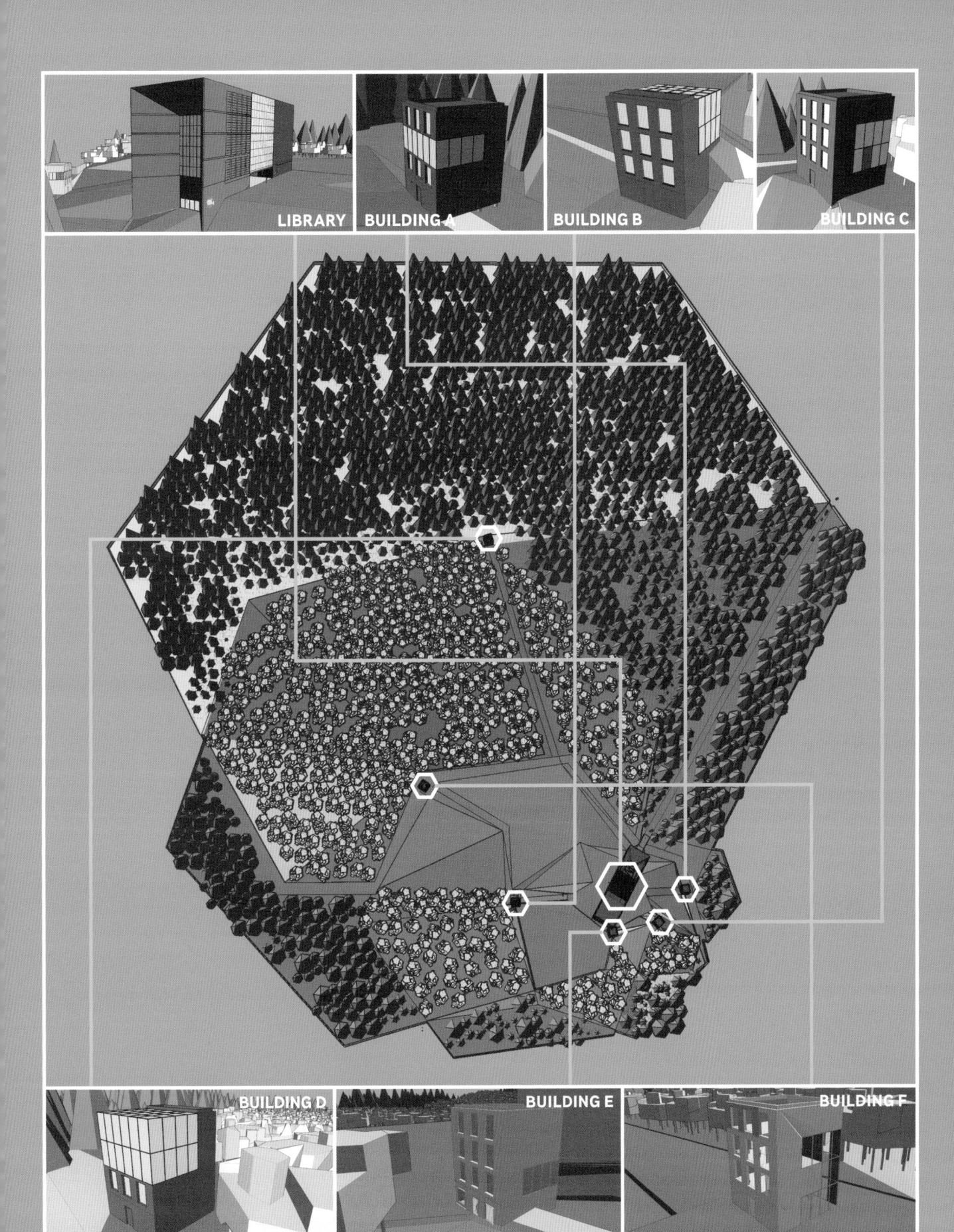

BUILDING D | BUILDING E | BUILDING F

Institute for the Study of Complex and Dynamic Systems
aerial view

Is it much farther?
Hello, Mr. Sorokin.
Hello.
All the preparations are in place. The Institute is at your disposal. You'll find any info you need in your office.

C'mon, boys. Rolling out!
Hey, Sorokin. Try to handle things better than the guy before you. It was a disaster area when we showed up.
I'll do my best.

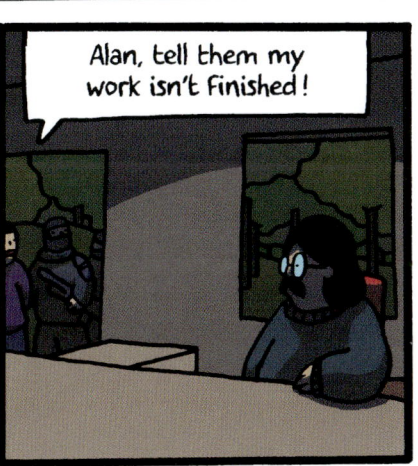

Tap!

Hello. I'm Alan Bateson, a researcher in systems sociology, and the Director of Institute no. 3.

If you're watching this video, it means that you were selected for the program just like I was, seven years ago.

You're the first resident and thus, the Director of Institute no. 4.

As I speak, Institute no. 3 is coming to an end.

No! I refuse to leave!

Alan, tell them my work isn't finished!

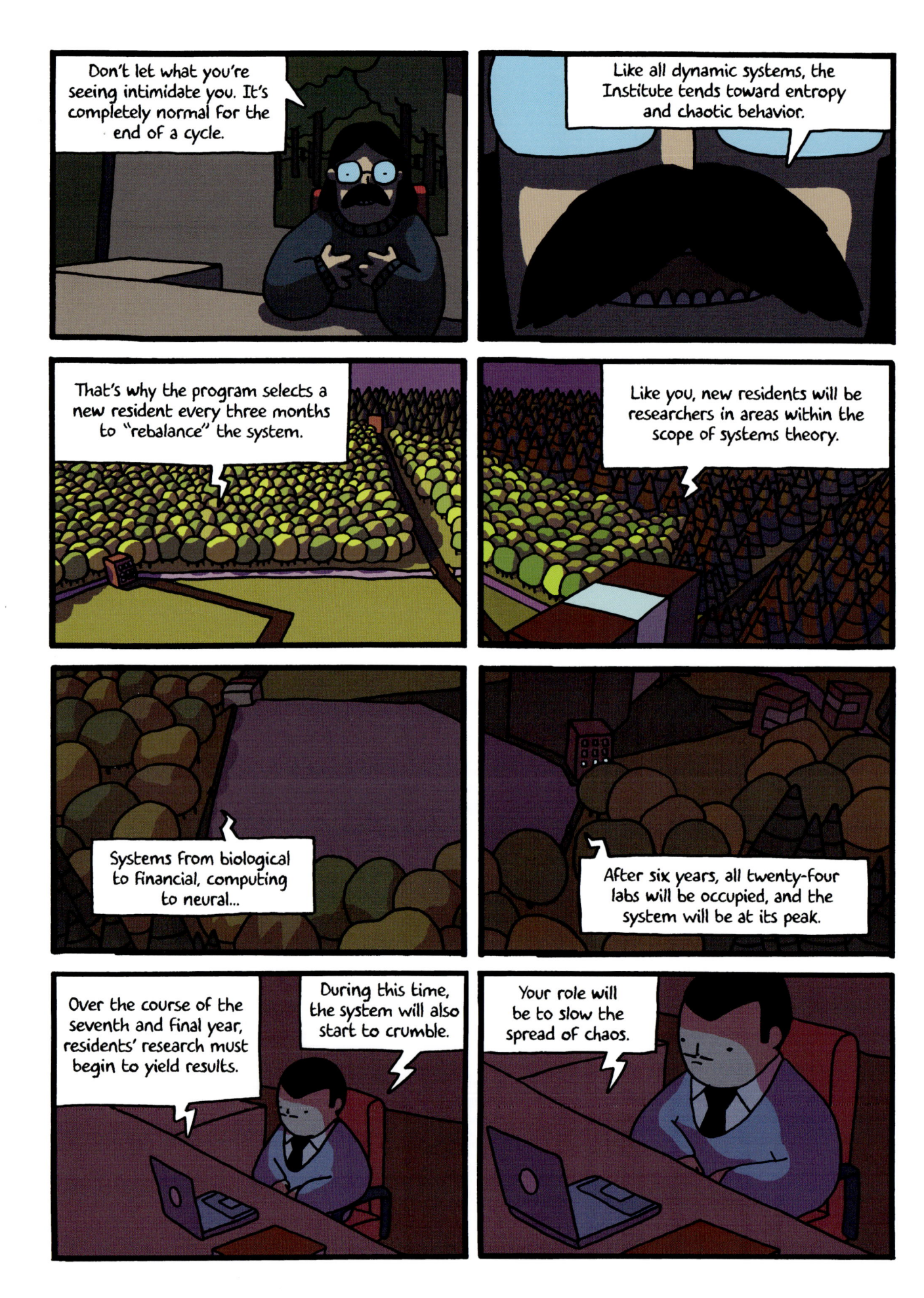

Don't let what you're seeing intimidate you. It's completely normal for the end of a cycle.

Like all dynamic systems, the Institute tends toward entropy and chaotic behavior.

That's why the program selects a new resident every three months to "rebalance" the system.

Like you, new residents will be researchers in areas within the scope of systems theory.

Systems from biological to financial, computing to neural...

After six years, all twenty-four labs will be occupied, and the system will be at its peak.

Over the course of the seventh and final year, residents' research must begin to yield results.

During this time, the system will also start to crumble.

Your role will be to slow the spread of chaos.

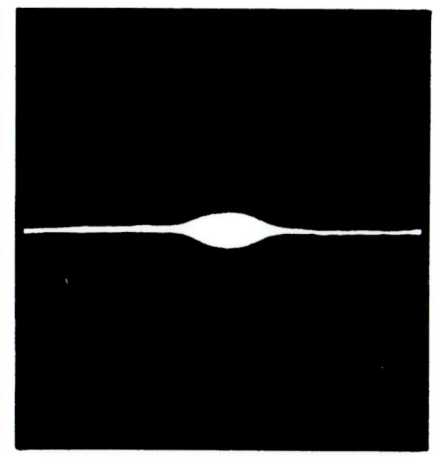

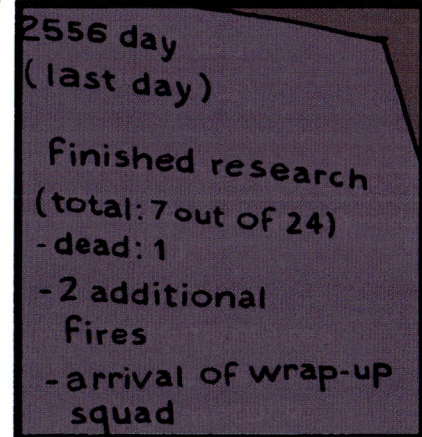

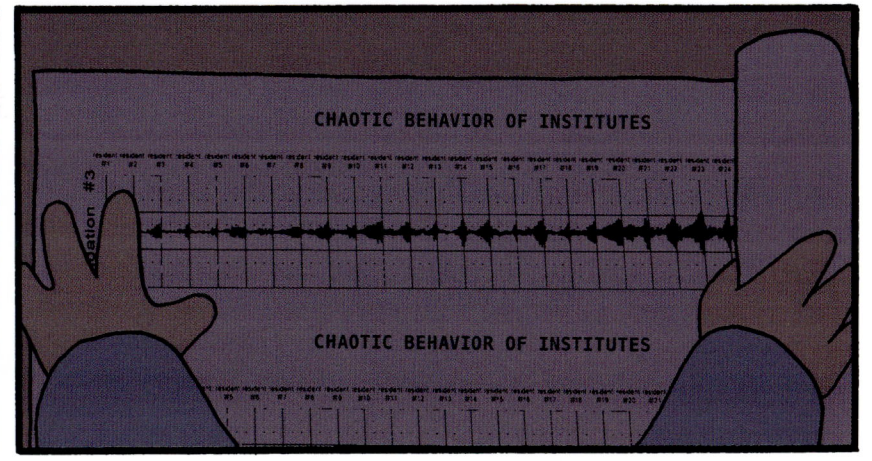

Panel 1: Hour's up, sir. Please come with us.

Panel 2: Just a sec.

Panel 3: Your role is to guide the residents but also improve the program.

Panel 4: This is the sum total of my results. / C'mon, load him up.

Panel 5: Wait...

Panel 8: CLAP

Panel 9 (text):
2556 day (last day)

- finished research (total: 7 out of 24)
- dead: 1
- 2 additional fires
- arrival of wrap-up squad

Panel 11: CHAOTIC BEHAVIOR OF INSTITUTES

CHAOTIC BEHAVIOR OF INSTITUTES

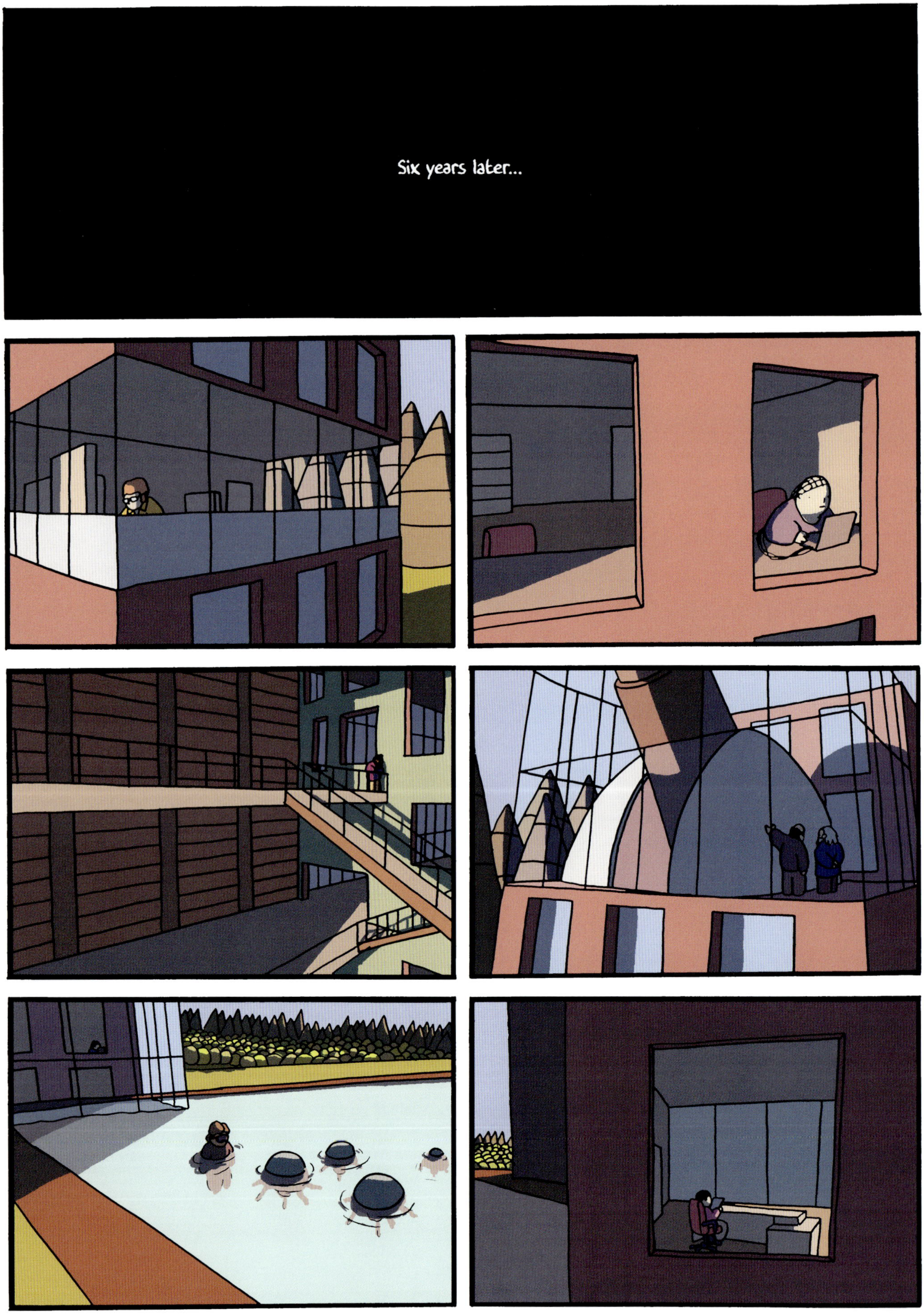
Six years later...

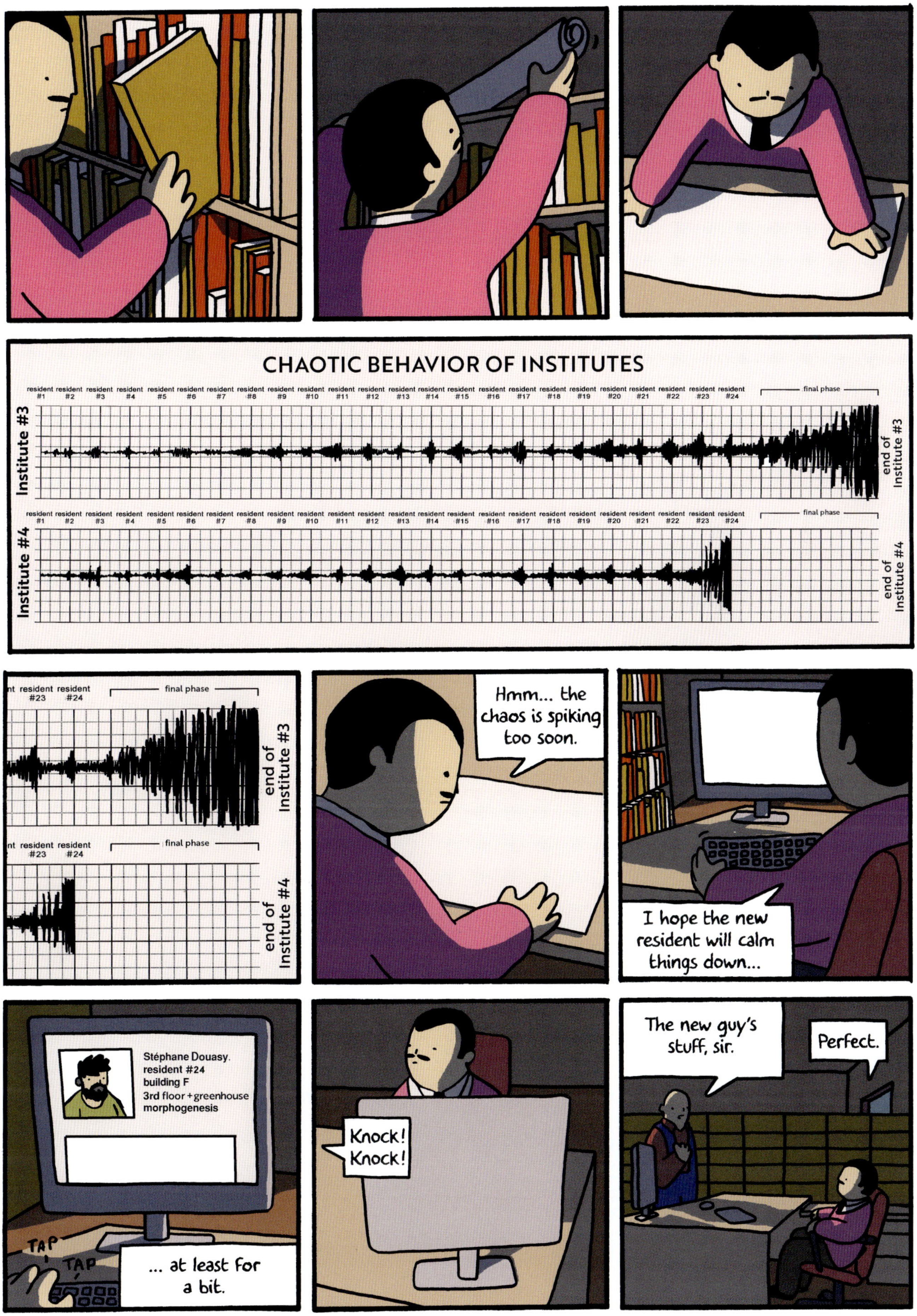

CHAOTIC BEHAVIOR OF INSTITUTES
resident #1 resident #2 resident #3 resident #4 resident #5 resident #6 resident #7 resident #8 resident #9 resident #10 resident #11 resident #12 resident #13 resident #14 resident #15 resident #16 resident #17 resident #18 resident #19 resident #20 resident #21 resident #22 resident #23 resident #24
final phase
Institute #3
end of Institute #3
Institute #4
end of Institute #4
Hmm... the chaos is spiking too soon.
I hope the new resident will calm things down...
Stéphane Douasy.
resident #24
building F
3rd floor + greenhouse
morphogenesis
TAP TAP
... at least for a bit.
Knock! Knock!
The new guy's stuff, sir.
Perfect.

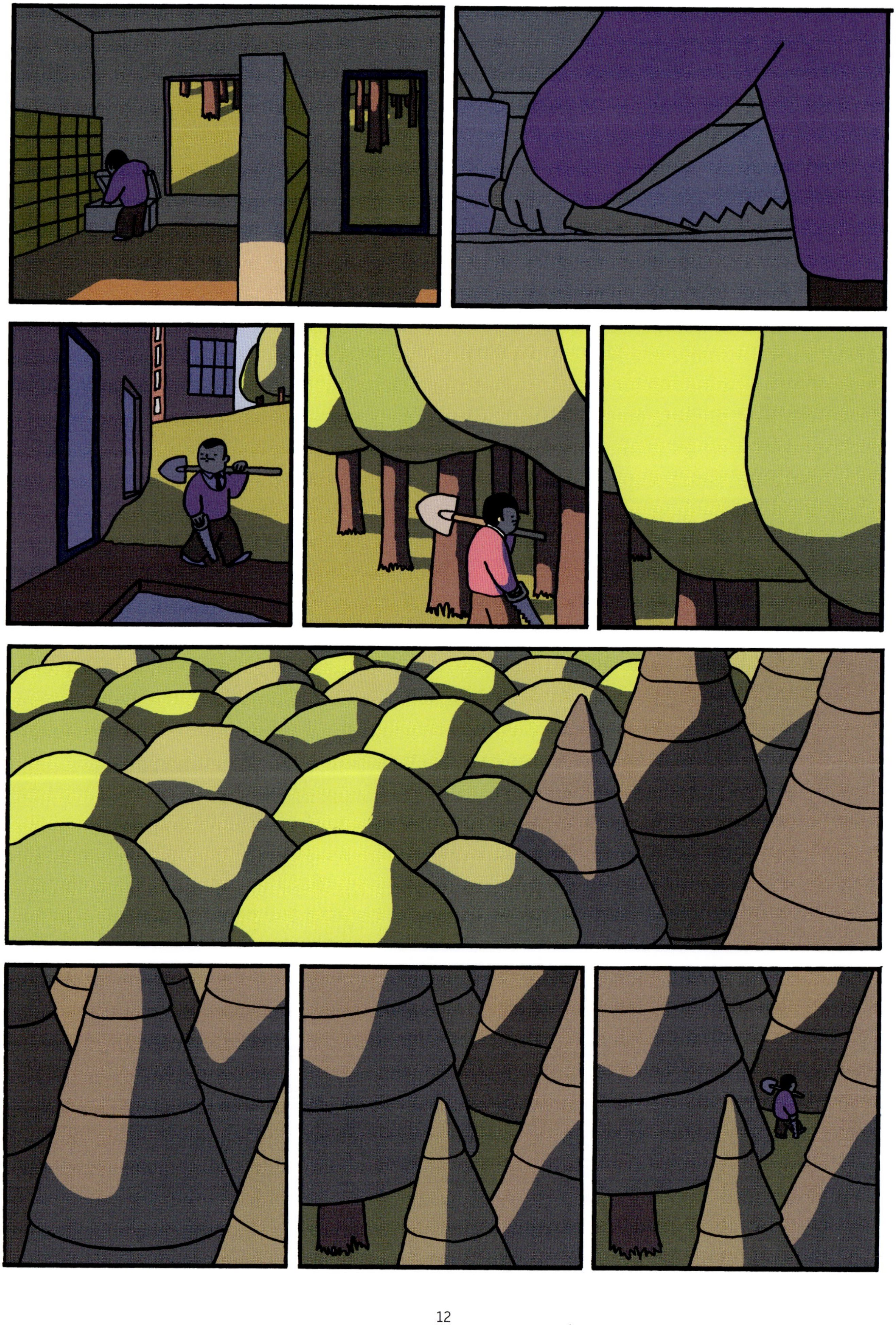

Philippe! Come look!

In the woods. It's happening again! And now, with smoke!

So what?

So what? It's weird, is what. Wonder what they're up to...

None of our business. Come on, back to work!

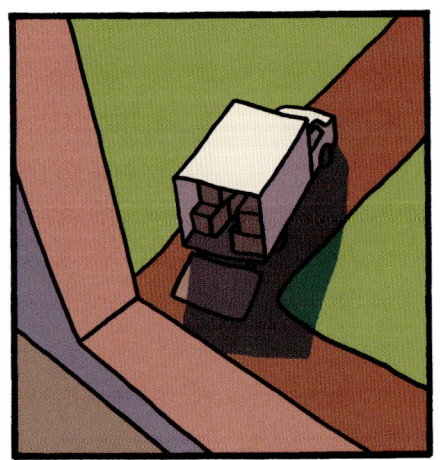

03:22:25 prob.
subject cold from
hexagonal and cat
reddish

03:22:34 prob.1
Death of subject

09h07m
16s20c

Tzz
tzzzg!

Crap! New
blood!

Building E...

Come in, come in!

Ah, Mr. Douasy. How was your trip?

Fine.

Good.

Looks like your lab's in Building F, third floor.

Your keys.

For food and daily supplies, please see the Logistics Manager.	If you need scientific equipment or encounter problems with your research, come see me directly.	All right.

The movers are setting up your things.

Your plants will arrive tomorrow.

Any questions?

Uh... no, not for now.

You are free to conduct your research as you see fit.

We ask only that you file regular reports.

As you know, you are the twenty-fourth and final researcher. That makes you a very special resident.

I'm counting a lot on your presence in our scientific community.

Uh... Thanks. That's nice.

Oh! Almost forgot. Any major or irreversible problems, here's the number of our cleaning crew.

Can I get to my lab this way?

Yes. Just go around the lake.

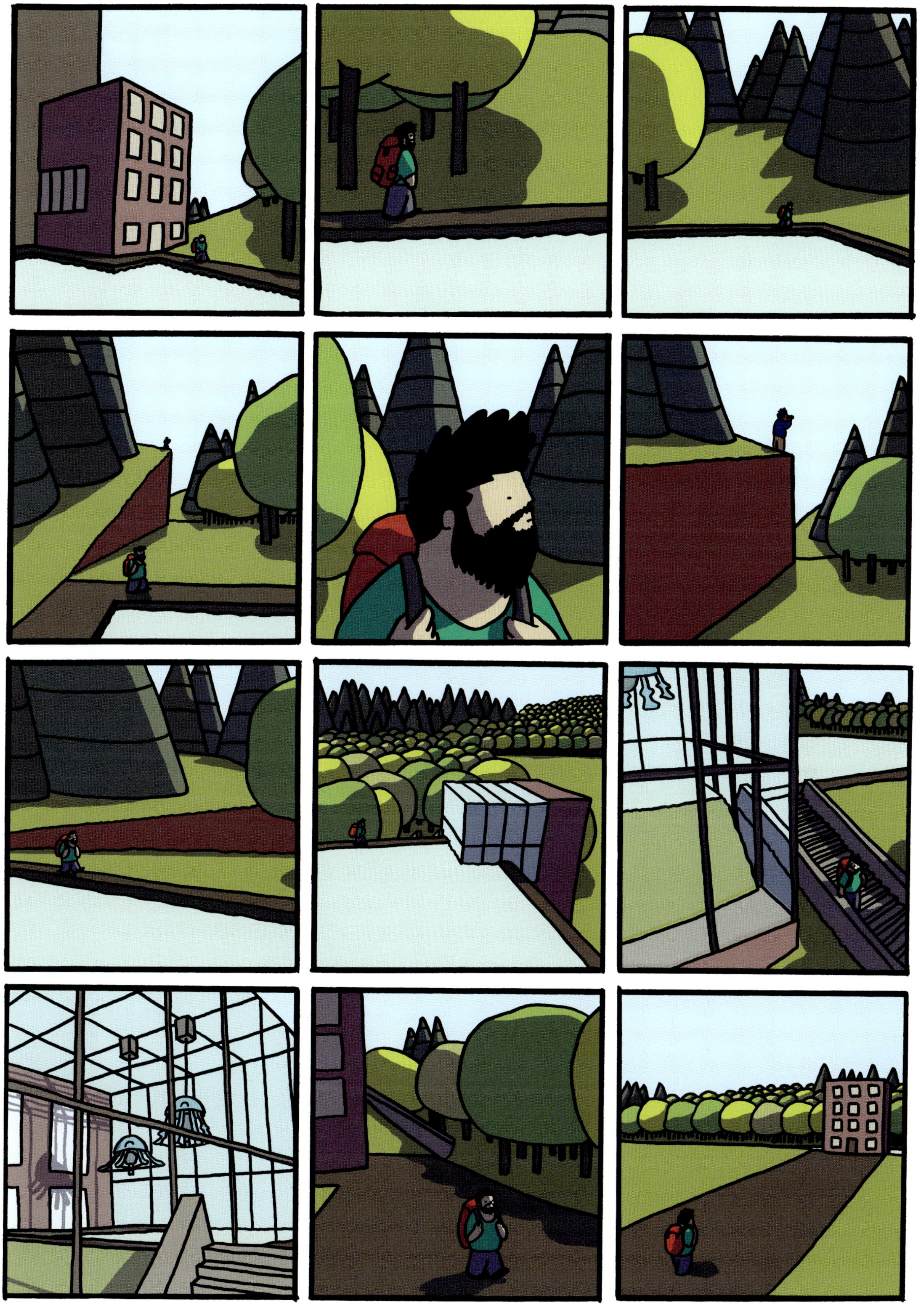

Whap!
Floom!
Aw, no more cigs ?!
Hello!
Sorry, we're full up!
No worries, I'll take the stairs.
Hiya!
Hi.
You the new one upstairs?
Yup. Third floor.
I'm Louise.
Stéphane.
Can I bum a smoke off you??
Oh, no, sorry.
I've got loose-leaf upstairs if you roll.
Cool. I can get a look at your lab.

This place is deluxe!
Whoa... you've barely unpacked yet!
yeah.
I've got some tea fixins somewhere...
No thanks.
So what's your field?
Morphogenesis. You?
Linguistics. But I'm not doing much right now.
Hey, Michel! Pallet jack giving you grief?
Pfft!
Oh, I see. They're using dolly straps.
OK, wimp?
Agh!
Looks like a breeze for them. Crazy.
Check out their posture. They've got technique.
You know the sofa problem?
"What's the largest area of a sofa that can be maneuvered...
"... through an L-shaped hallway one meter wide?"
No one's ever solved it, but I'll bet a mover could eyeball it in an instant.
Funny... I'm in an irreversible system, and yet I'm managing to advance.
Whoops! And it stopped.
Knock! Knock!
Hello.
You the new guy?
Yep.
Well, I'm off. Good luck!
Hi, Vilhem.
Hello, Louise.

Why wasn't I informed of your arrival?
Uh...
Did you fill out your personal file?
Right here. Wasn't sure whom to give it to.
And you! Will you be much longer?
I'm done...
Perfect.
Sorry to disturb you.
No worries.
Klik.
Whoa!

You know you're to notify me of the slightest change to the Institute's daily routine!
Especially in Building F, where my experiment is taking place!
Don't tell me my job, Vilhem. Everything's been planned out. I was not to warn you of Mr. Douasy's arrival.
Had I done so, you would have holed up...
... and never spoken a word to him.
Your so-called "research" cost me a week of work! Was that part of your plan?
Anything else I should know?
Mr. Douasy's plants will be delivered tomorrow.
Great. No modeling for two days! I want a list of those plants ASAP.
Your models have yielded nothing in six years!
New entry
-Stephane Douasy
-living creature
-human
-male
-32 years
-Building F
TAP!
TAP!
NAME:
SURNAME:
AGE: 32

Knock! Knock!
Hello hello.
Hello.
Sleep well?
Terrific. With all these trees, it's like being in the woods.
C'mon, I'll give you a tour.
Uh...
Don't argue. You can unpack later.
Ah! Must be for you.
No way! They bought all of it!
We get whatever we want. You'll get used to it.
That, Stéphane, is your lab. It's the nicest: all those windows!
It's located in Building F, which, as you can see...
...is shaped like an F.
It's all logical around here. Ever since the last Institute. Even the forest is planted according to an anti-fire algorithm.
Insane, right?
Every building looks like a lowercase letter: a, b, c, d, e, and f. The proportions of the Institute's layout follow a sequence...
Fibonacci.
You noticed?
1, 1, 2, 3, 5, 8, 13, 21, 34, 55, 89, 144... I use it all the time.
This is Lab B.
Oh right, I saw their robo-medusa thingies.

"This is Building E, for research on Institute management and optimization." / "Wait, they research that, too?"	"Everything here is subject to scientific research." / "The Institute itself is an experiment in systems sociology."	"You met Martin Sorokin. He's continuing the experiment of the first three Institutes. He's kind of... dedicated, you could say."
"Our massive financing comes from research on the mechanics of financial systems."	"Another roboticist, over there."	"Oh right, I forgot! The library!"
"The Institute's nerve center, and the only place you might run into another resident. Otherwise, it's a ghost town."	"Astrophysicists over there. Never see 'em." / "So then... what's daily life like?"	
"It's a perfectly oiled machine. Not a cog out of place. But as far as a social life..."	"Let's just say you're the first person I've talked to in two weeks."	"I'm very happy to have you as my new neighbor."

Here's a list of the plants. I added the plans for the work we did.
Perfect, thanks.
New entry:
-Plant 1
-living creature
-plant
-Aloe polyphylla
-5 years
-building F
-greenhouse
Green-house is done.
I'll check out the botanists' work. See you...
Ciao.

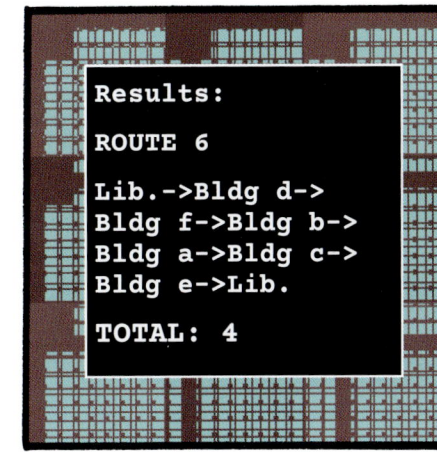

DISTANCE: 53.01m

ROUTE #6:
Lib. → Bldg d: 37.22m
Bldg d → Bldg f: 66.02m
Bldg f → Bldg b: 162.17m
Bldg b → Bldg a: 06.03m
Bldg a → Bldg c: 51.15m
Bldg c → Bldg e: 42.18m
Bldg e → Lib.: 53.01m
TOTAL: 407.73m

```
Results:
ROUTE 6

Lib.->Bldg d->
Bldg f->Bldg b->
Bldg a->Bldg c->
Bldg e->Lib.

TOTAL: 4
```

Wait... it works?

I—it can't be. There must be some mistake!

E3
FINANC
STOCK
E1
Leave Paniandy's supplies in front of the door. He'll pick them up tonight.
He's not a wild animal!
Just do as I say!
OK, OK.
Hello Stéphane.
Hello.

"Mr. Douasy, right?"

"So, how does resupply work?"

"Simple: you tell me what you want, and I give it to you."

"Got any duck breast?"

"Yes."

"How 'bout wine?"

"Ho ho! Follow me!"

KLINK! KLINK!

KNOCK! KNOCK!

"Helloooo?"

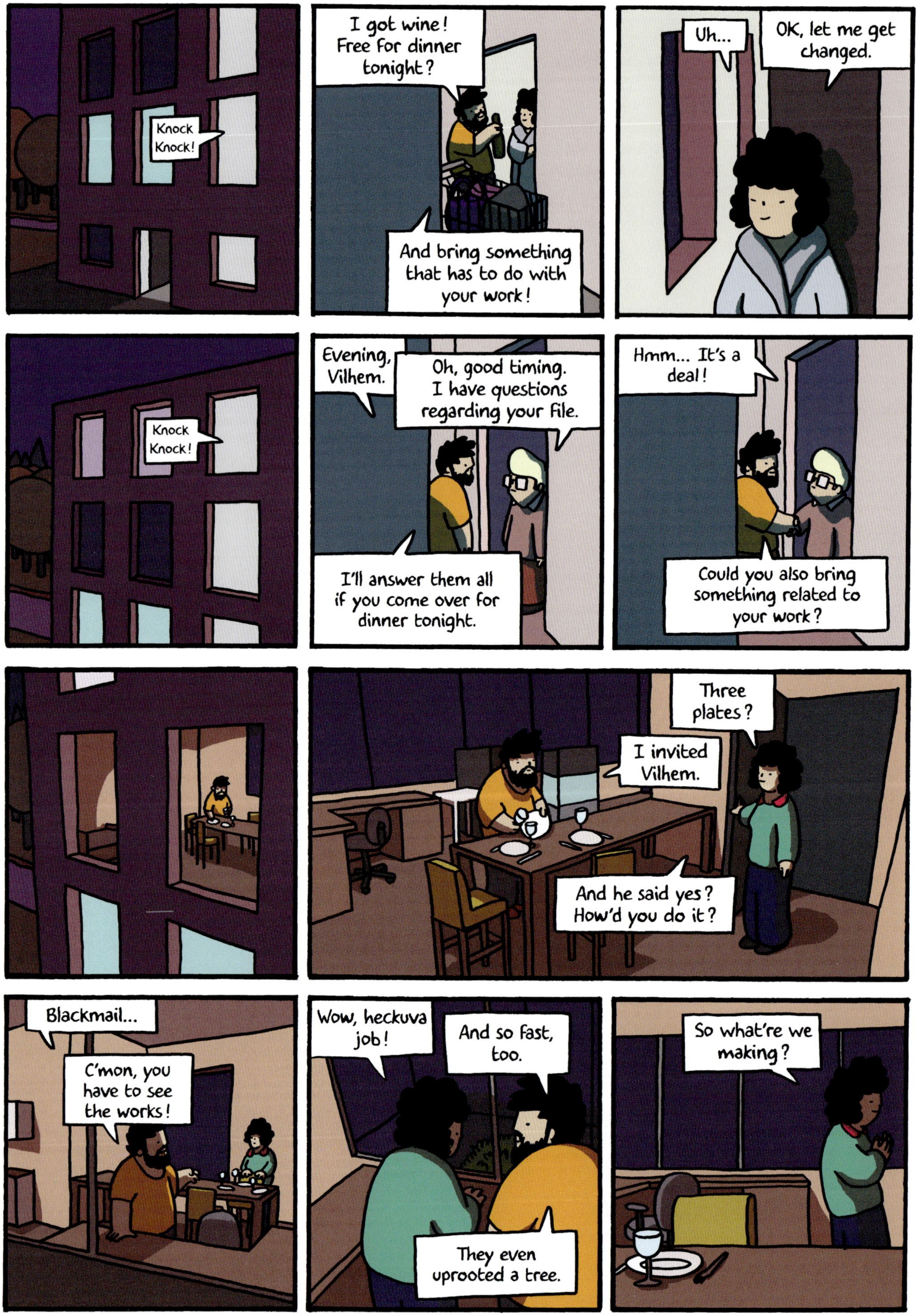
Knock Knock!
I got wine! Free for dinner tonight?
And bring something that has to do with your work!
Uh...
OK, let me get changed.
Knock Knock!
Evening, Vilhem.
Oh, good timing. I have questions regarding your file.
I'll answer them all if you come over for dinner tonight.
Hmm... It's a deal!
Could you also bring something related to your work?
Three plates?
I invited Vilhem.
And he said yes? How'd you do it?
Blackmail...
C'mon, you have to see the works!
Wow, heckuva job!
And so fast, too.
They even uprooted a tree.
So what're we making?

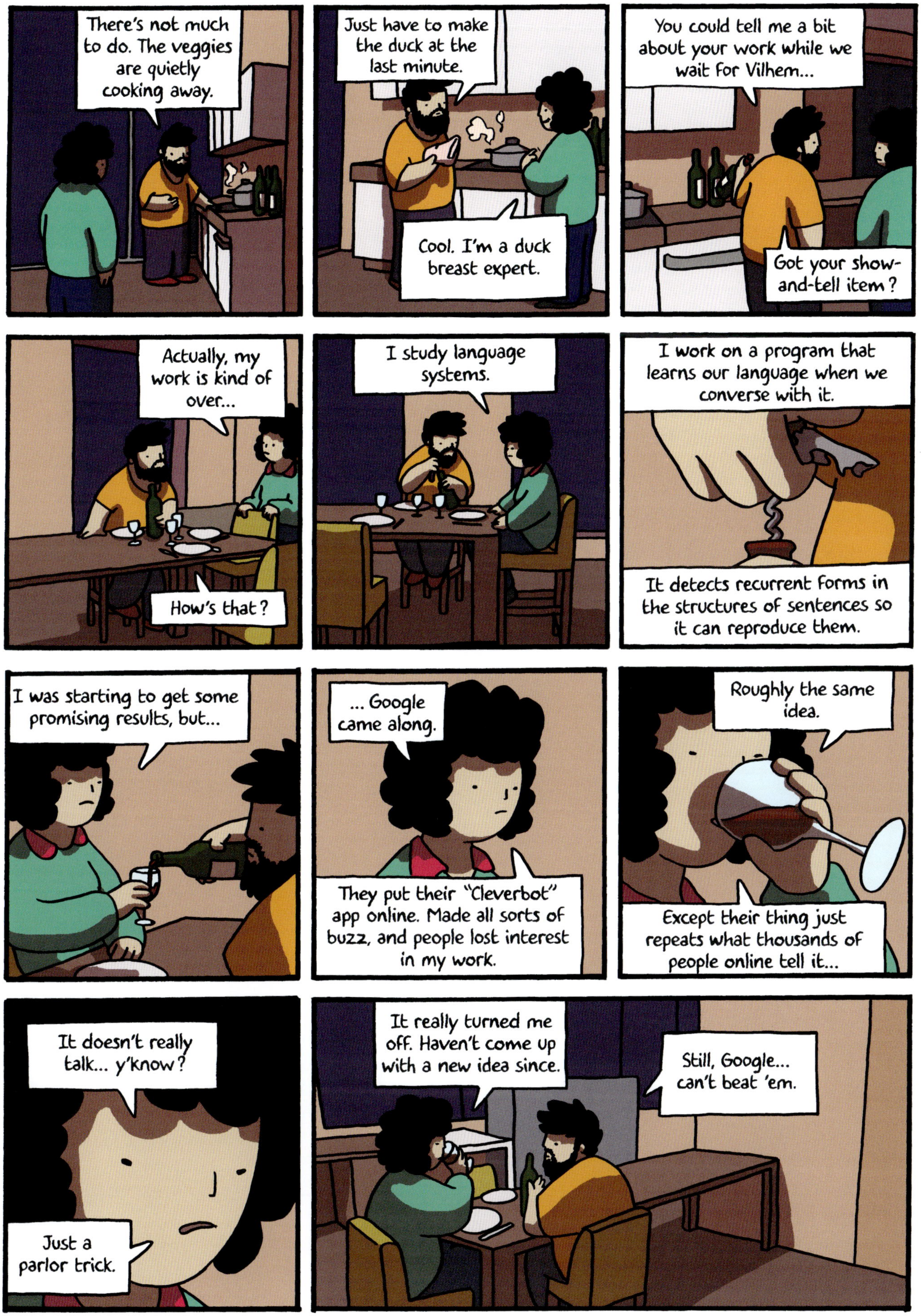
There's not much to do. The veggies are quietly cooking away.
Just have to make the duck at the last minute.
Cool. I'm a duck breast expert.
You could tell me a bit about your work while we wait for Vilhem...
Got your show-and-tell item?
Actually, my work is kind of over...
How's that?
I study language systems.
I work on a program that learns our language when we converse with it.
It detects recurrent forms in the structures of sentences so it can reproduce them.
I was starting to get some promising results, but...
... Google came along.
They put their "Cleverbot" app online. Made all sorts of buzz, and people lost interest in my work.
Roughly the same idea.
Except their thing just repeats what thousands of people online tell it...
It doesn't really talk... y'know?
Just a parlor trick.
It really turned me off. Haven't come up with a new idea since.
Still, Google... can't beat 'em.

Their model's super well-established. They're covering all the bases.
Still, what do you care? Keep on truckin' in your own little corner. Improve your app and write a real paper!
I don't feel like it anymore. It wouldn't be new. They got there first.
That's what drives me nuts, see?
What?
Having the same idea as those lunkheads!
Do you get a lot of pressure around here to find a new topic?
From Martin, you mean? I tell him, "It's going."
But I can't take forever...
Bam Bam!

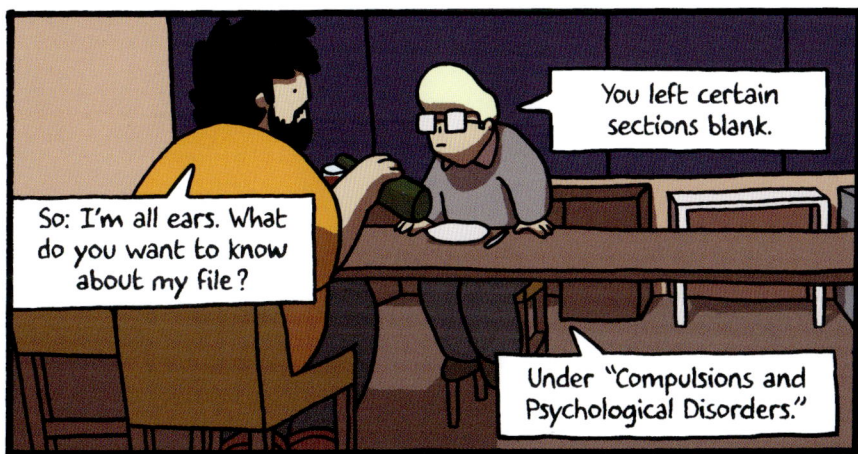

— Ah, Vilhem! We were just having cocktails.
— Oh...
— Call you Vill?
— Uh... Ok.

— How 'bout a glass of wine? It's delish.
— Just one.

— I'll start the duck?
— Yell if you need a hand.

— So: I'm all ears. What do you want to know about my file?
— You left certain sections blank. Under "Compulsions and Psychological Disorders."

— Look, we all have compulsions. I, for one, can be a bit... fastidious.
— Ha ha ha! A bit!
— Don't think I have any. Never seen a shrink, if that's your question.

— Just what do you do, that you need info like that?

— He thinks he's Nostradamus!
— Hmf.

— I'm working on a computer program to predict my actions in the near future.
— Duck's ready.

— A program that predicts your future? Sounds like Sci-Fi.
— A matter of perspective.

— Chaos theory clearly explains why you can't predict the future.
— Every event is the result of an infinite number of causes. There's no way to observe them all.

— In theory, the future can be predicted, but in practice, it's unfeasible.
— Tell that to a meteorologist.

— What I'm doing is like that.
— Looking for the scenarios likeliest to occur.
— I take only the largest parameters into account.
— Such as?

— First of all, personal parameters: my childhood memories, my physical features, my routine...

— Even this dinner's going right into his program!

— Next, environmental parameters: the Institute, its residents. Including you...

— Ok, now I get all the weird questions.

— You'd need a really powerful computer to handle all that data...

— I have a supercomputer.
— Whoa!

He forgot to mention that his program hasn't yielded any results in six years.
Six?
He's the second resident, after Martin Sorokin.
In fact, there's been progress.
Look...
I used to only get results like this:
13:47:11
Prob.: 38%
Choosing them of fluid rights itself
13:47:11
Prob.: 38%
Choosing them of fluid rights itself of subject
But recently, I managed to get a real sentence.
13:58:12
Prob.: 11%
SUBJECT TRANSFER ITSELF 22° WEST
That makes no sense!
You're the subject?
Yes.
Now that the program's working, I just have to get it to give correct results.
Why does it say "Death of subject" at the end?
That's how the reports always end.
Must be a bug...
Kinda freaky.

And what are you up to with those plants?

What are you, a biologist? I bet you study living systems.

He studies natural forms, not metabolisms.

I'm a physicist, not a biologist.

Here's my research, right here! Bon appétit!

Romanesco broccoli?

Never had it. You study fractals?

No, I study the origin of forms in general.

Right now, I'm analyzing vegetable forms.

For instance, why asparagus and Romanesco broccoli are organized, like most plants, in spirals...

Ah, indeed. Fascinating.

And beautiful.

Whoa, I'm stuffed.
And tipsy, too!
This wine is indeed tasty.
So... there's no one on the ground floor?
I knocked earlier. No answer.
Of course there is...
The Phantom Scientist!
What?
No. There's no one.
Officially, there are now four of us in Building F.
But we've never seen nor heard the fourth.
Surely a bureaucratic error.
Or a resident who never showed up...
Or else he's veeery quiet.
Only one way to find out for sure!

Whoa... feeling a bit tight.
It's so late! Must be asleep...
BAM! BAM!
Anyone home?
Hee!
Shhh!
Hey!
BAM! BAM!
Can either of you pick a lock?
No need. Got the keys...
Not this one...
KLIK KLIK
Why do you have the keys?
I have all the keys!
KLIK. KLIK.
KLAK!
Ah.

"It's *not* empty!"

"Dirty dishes in the sink."

"Oh, crap!"

"We should go."

"Great. Another resident!"

"That's going to add processing time!"

"Well, we learned something today."

"Now we know who's got the worst lab!"

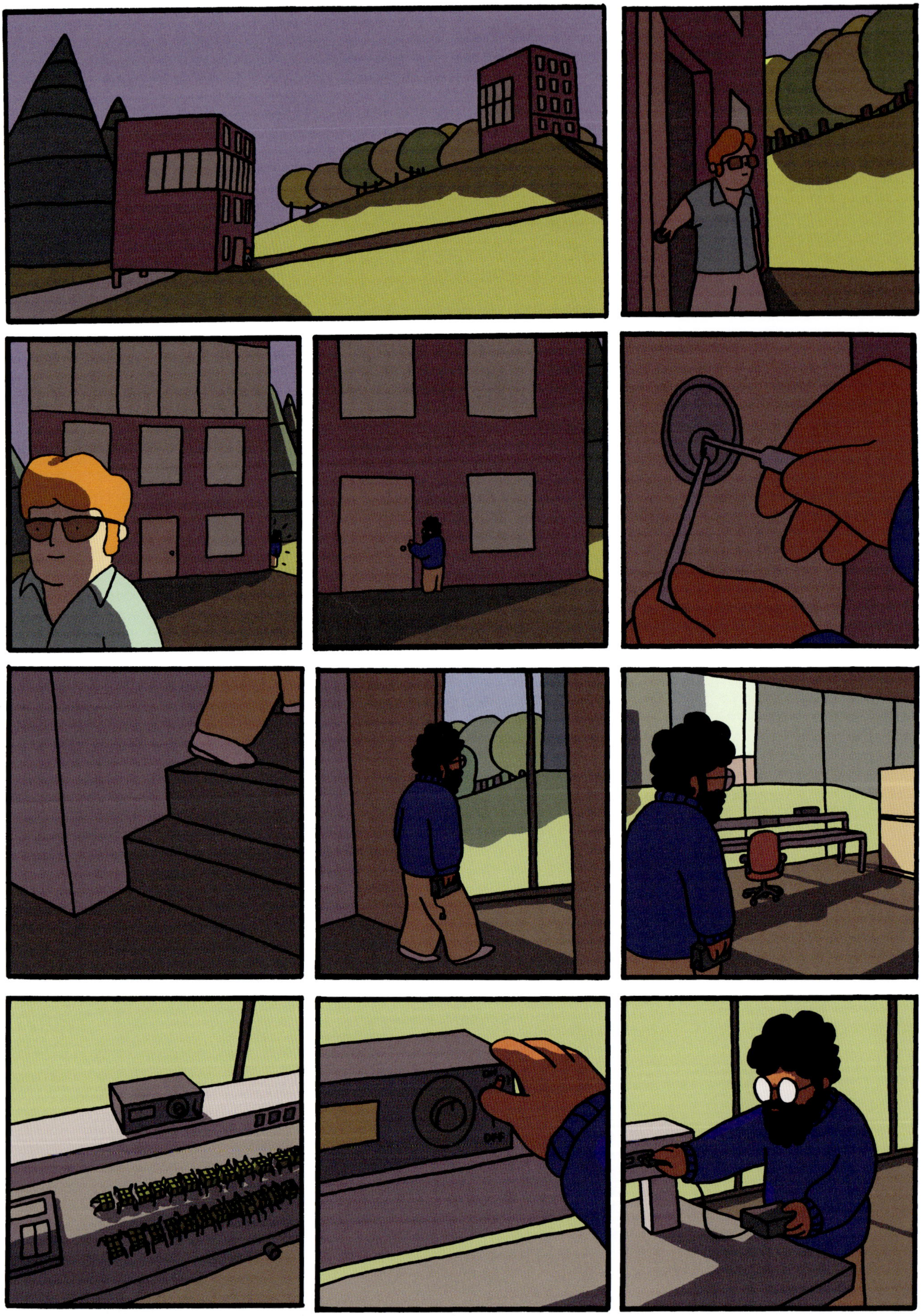

"Hello hello!"

"It obviously doesn't work, Chizuru!"

"But I'm sure they can see you!"

"I don't understand why you're still working with those old robots."

"They're more durable and reliable than yours!"

"What can we do for you, Han?"

"Just dropped by to say hi."

"Can't you see we're busy?"

"Keep your advice to yourself. Do I go over there and tell you how to make your doohickeys?"

"Gotta know when to take a break now and then."

But robots are kind of my department...

Nothing personal, but we're not even talking the same league here.

No need to be a jerk about it!

There's a teensy difference between DIY engineering and the product of rigorous scientific research.

I...

You're just a simple maker. We're *artistes*.

You still needed this "simple maker" to perfect your robots, or you'd still be having technical difficulties.

Pretentious ass!

Quit looking at me like that!

Let's see, what was it again...
Some kind of Indian name.
Ramia... Pamia...
Aha! Paniandy.
Great. File's virtually empty.
TAP
TAP
Knock! Knock!
TAP
TAP
Come in!

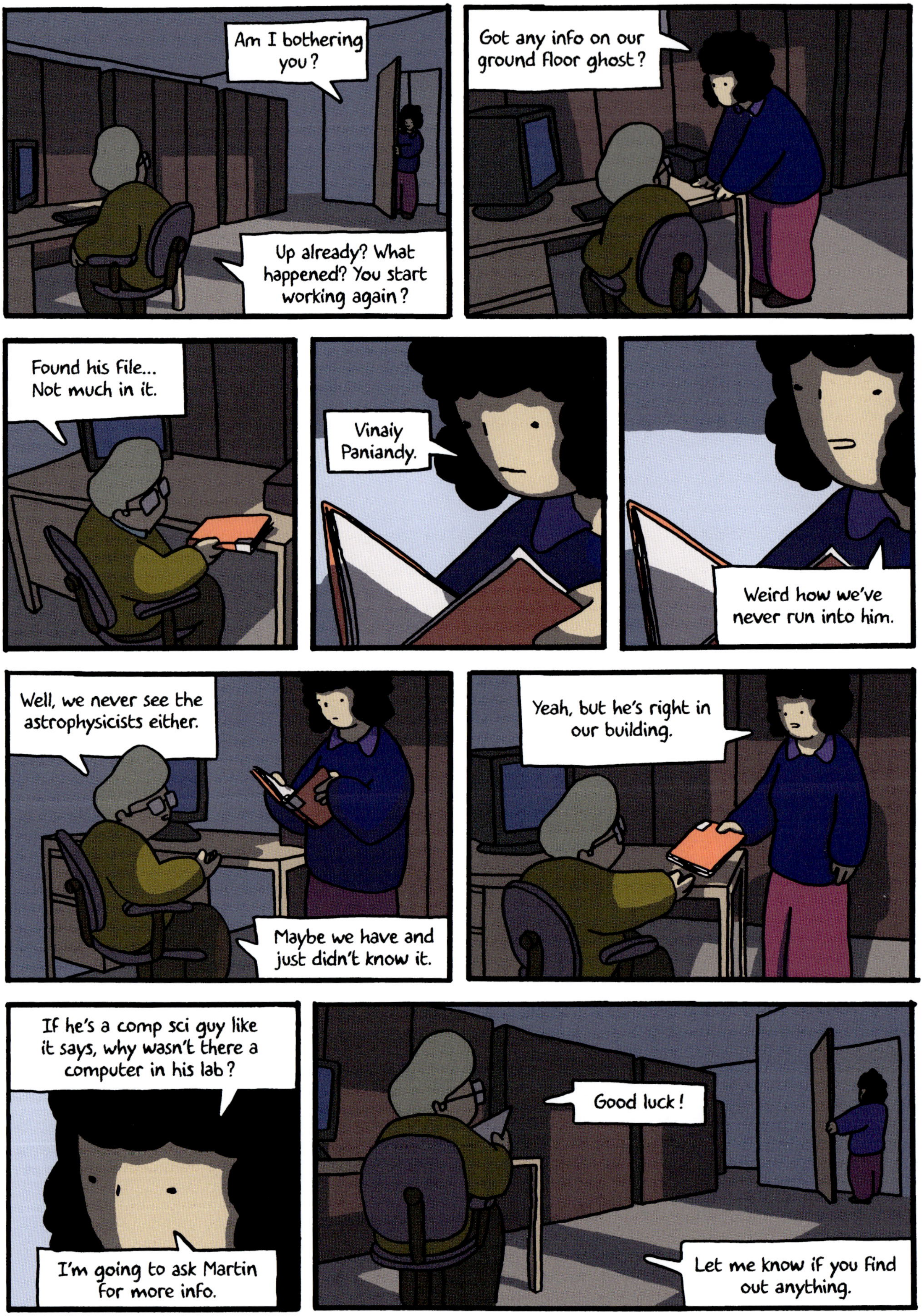
Am I bothering you?
Up already? What happened? You start working again?
Got any info on our ground floor ghost?
Found his file... Not much in it.
Vinaiy Paniandy.
Weird how we've never run into him.
Well, we never see the astrophysicists either.
Maybe we have and just didn't know it.
Yeah, but he's right in our building.
If he's a comp sci guy like it says, why wasn't there a computer in his lab?
I'm going to ask Martin for more info.
Good luck!
Let me know if you find out anything.

"Sorry, I can't give you any information on Vinaiy Paniandy."

"You could say if you've met him, or what he looks like."

"No. That would be detrimental to the Institute's operation."

"That's ridiculous!"

"My results brook no argument."

"No information on Vinaiy Paniandy can be shared with the Institute's other residents."

"That would—"

"be 'detrimental to the Institute's operation.' Spare me, Martin."

"Is he even still here?"

"I cannot say."

"You've become unbearable, Martin. Acting the dictator."

"You'd best save your energy for your research."

"Is it coming along?"

"Yes!"

Day 2198. 10:30 a.m.

KLIK

The Paniandy problem has resurfaced.

The recent quiet was but a pretense.

resident #23   resident #24

resident #23   resident #24   final phase
FIRE
DEATH OF RESIDENT #8
FIRE
end of Institute #3

resident #23   resident #24   final phase
PANIANDY PROBLEM!
end of Institute #4

		Foundation #3 Day 2198
"Day 2198, 9:15 p.m. It's all going according to plan."	"The chaos is under control. It seems that..."	"Hmph." *CLAP*
"Knock knock!" "What now?!"	"Dunno. You called for me!"	"Oh, right, uh... I need a sleeping pill."
"Look, I've told you before. I'm no doctor."	"Have a glass of whiskey. It'll chill you out."	"Bring me a sedative right now!"

45

Hello!

Hmph!

Bunch of weirdos!

Library

Search:

Vinaiy Paniandy

1 result(s)
- Articles (**1**):
- «**P≠NP**»

Floor: 2
Cat.: art.
Ref.: pnp-pan01

Vinaiy Paniandy: Computer scientist born in New Delhi in 1971.

In 2005, Vinaiy Paniandy published a proof of P ≠ NP that made a lot of waves. The scientific community quickly pointed out a series of errors that...

"In theoretical computer science, the problem P vs. NP remains unsolved."

"It is considered by many to be among the most important problems in the field, or even mathematics in general."

"It is one of the seven Millennium Prize Problems on the Clay Mathematics Institute's list, each of which carries a US$1,000,000 prize for the first correct solution."

Sunday, May 15th, 8:30PM—11:30PM:
-Subject had dinner with Stéphane Douasy and Louise François
-Location: Stéphane Douasy's lab
-Menu: Romanesco broccoli, duck breast, and asparagus +7 glasses of wine. Conversation revolved around the scientific research of

Sunday, May 15th, 11:45PM:
-Subject entered ground floor lab with Stéphane Douasy and Louise François

The lab seems to be inha by Vinaiy Paniandy

New entry:
Vinaiy Paniandy

Processing, please wait...
prediction of events to occur in 24:00:00

Processing, please wait...
prediction of events to occur in 23:59:57

"The calculations will never be done in time."

"Vilhem? What're you doing here?"

"You're never at the library."

48

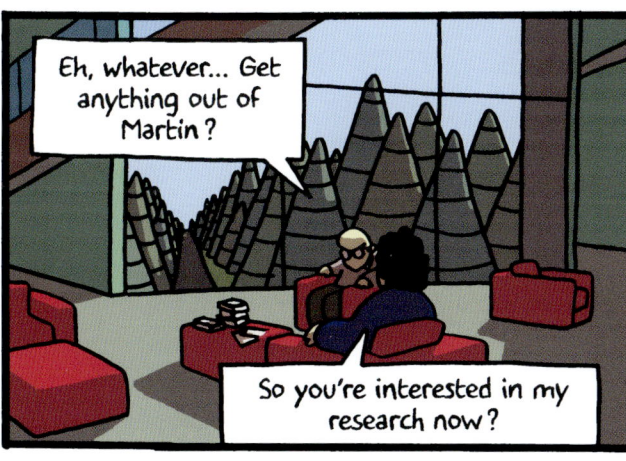

"I have nothing better to do today. I launched a prediction for 24 hours out, but it'll take a lot longer to process."	"So you can't predict it till after it happens. Not very helpful."
"Tell me about it! I have so little data on Paniandy that I'd be shocked if the prediction made any sense."	"At least you know someone lives there now. Your data will be more accurate."
"Eh, whatever... Get anything out of Martin? So you're interested in my research now?"	"Let's just say it's been a while since I've seen you work on anything."
"'P ≠ NP'... That ring a bell? Sure. It's THE comp sci problem... The grail!"	"Solving it would have major implications for modern computing and science as a whole... ...Even our perception of the world."

"Break it down?" / "Sure. It's a problem in algorithmic complexity theory..."	"What?" / "A method of classifying various scientific problems based on the capacity of computers to solve them."
"Some problems have already been proven to be undecidable. No program can ever solve them."	"Among decidable problems are ones whose solutions can be easily checked or verified: 'NP.' And others that can be easily solved, or 'P.'" / "We want to know if these two categories are really distinct..."
"How can we know to verify answers but not obtain them?"	"Imagine a puzzle... It's hard to solve: you have to try every piece."	"But when you're done, it makes a picture." / "Hard to solve, easy to check."
"If you've got instructions for where every piece goes, you just follow them, and you'll finish the puzzle much faster..." / "Without trial and error."	"In short, 'P' problems are ones where we have instructions." / "In comp sci, we call that an algorithm."

So then the question "Does P = NP?" comes down to: "Do all problems we can solve through trial and error necessarily have instructions?"

More or less...

What a nifty headscratcher!

So what are some paths to solving it?

Some NP problems combine the difficulties of all the other problems. They're called "NP-complete."

Find an algorithm that solves an NP-complete problem, and it'll work for all other NPs, and we'll have proven that NP problems are also P problems.

Thus, P = NP.

Conversely, if it's proven that no such algorithm exists, then P ≠ NP.

Any examples of NP-complete problems?

Um... Does this sudoku have a solution? Does this minesweeper board?...

Or take Tetris. Can you make over 150 lines with this series of pieces? These are all the same question...

All it takes to solve "P vs. NP" is finishing a game of sudoku?

Yes, but without guessing.

BIP
BIP
What are you printing?
At least it's not sudoku games... I have to go.
Wait...
I found Paniandy's blog. He posted a whole series of proofs that P ≠ NP.
Seventeen, to be exact.
Can you help me carry these?
Here, you take five journals, two newspapers, and... how 'bout these two doorstoppers.
Equal division of labor.
See? Another NP-complete problem!

"So what do you think? P = NP, or P ≠ NP?"	"I don't like prognosticating. Most mathematicians believe P ≠ NP." "That sets a theoretical limit on computing capacity."
"What if an algorithm for solving an NP-complete problem is found?" "If P = NP?"	"That would shake the very foundations of how mathematicians view the world..." "Such an algorithm could solve any problem that has a valid solution."
"Show it a finished sudoku, and it could solve all sudokus?" "Exactly."	"No way!" "Everything that could be coded in computing language and turned into an NP-complete problem could be solved." "Yes. Imagine showing it any old theory and its proof."
"The algorithm could 'analyze' the mechanism of the proof..." "...and prove theories, discover new theorems..." "It could get creative..."	"Whoa, That's amazing!"

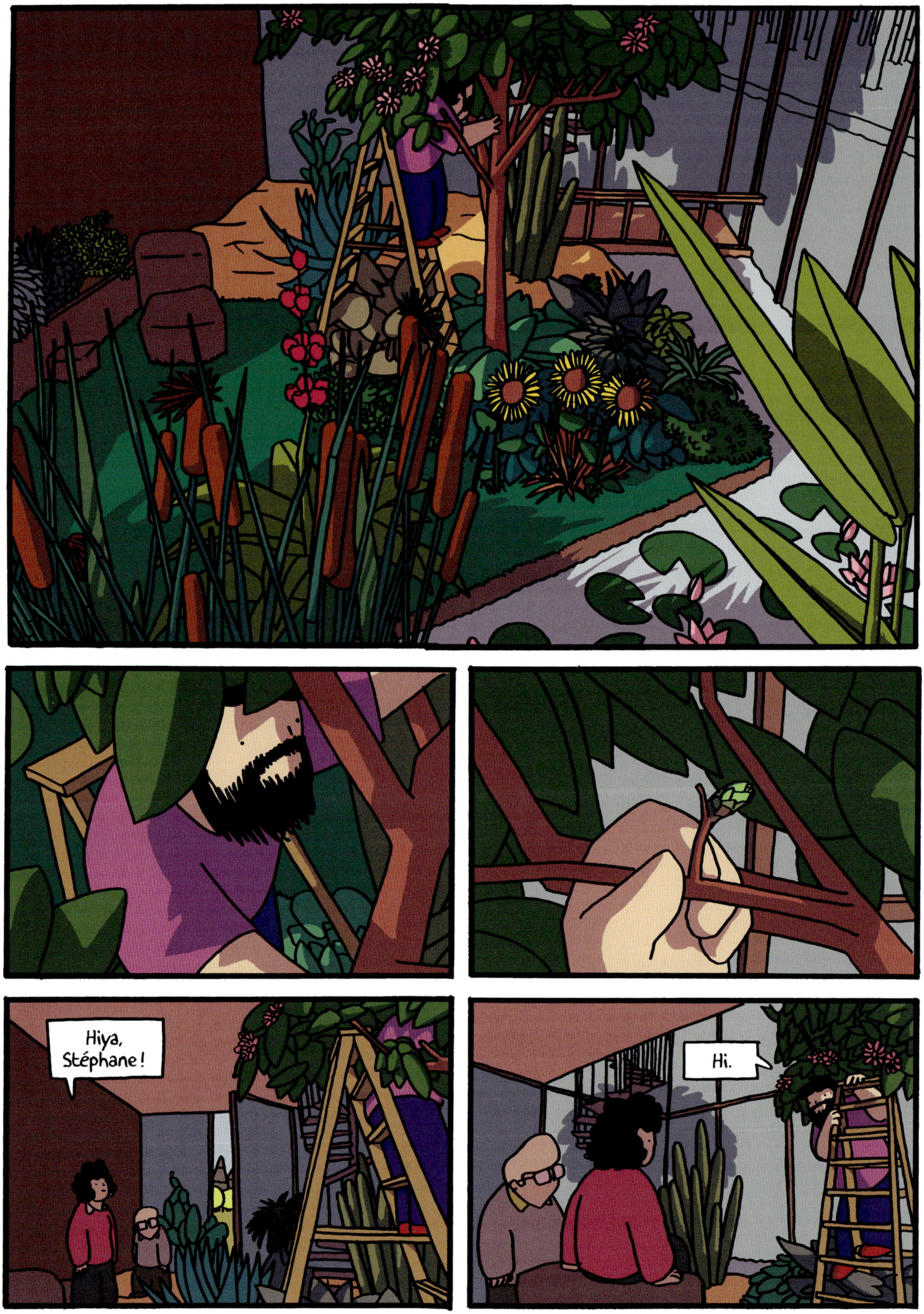
Hiya, Stéphane!
Hi.

55

— So... track down your ghost yet?

— Not yet. But we know what he's working on.

— I'll be going now.

— By the way, I updated my file. Will that work for you?

— Yes, I saw it. Bye.

— What's his deal?

— His program's even slower now. Too much data to process.

— Oh...

— Don't you want to hear about our neighbor?

— Sure.

— Well, guess what?

| ... and apparently it's one of the biggest problems in modern mathematics! | Paniandy thought he solved it? | Yeah, but people found three logic errors in his article... Poor guy. Imagine! |

| After that, he came to the Institute. | Like my grandmother said: "Patience and perseverance are the mothers of science." |

| Yeah. She was right. | Now I'm intrigued, see? I absolutely must investigate. | I'll grab Vilhem's keys and wait for Paniandy at his place. |

| He'll show up sooner or later! | How's your work going? | I might have a lead on buds. |

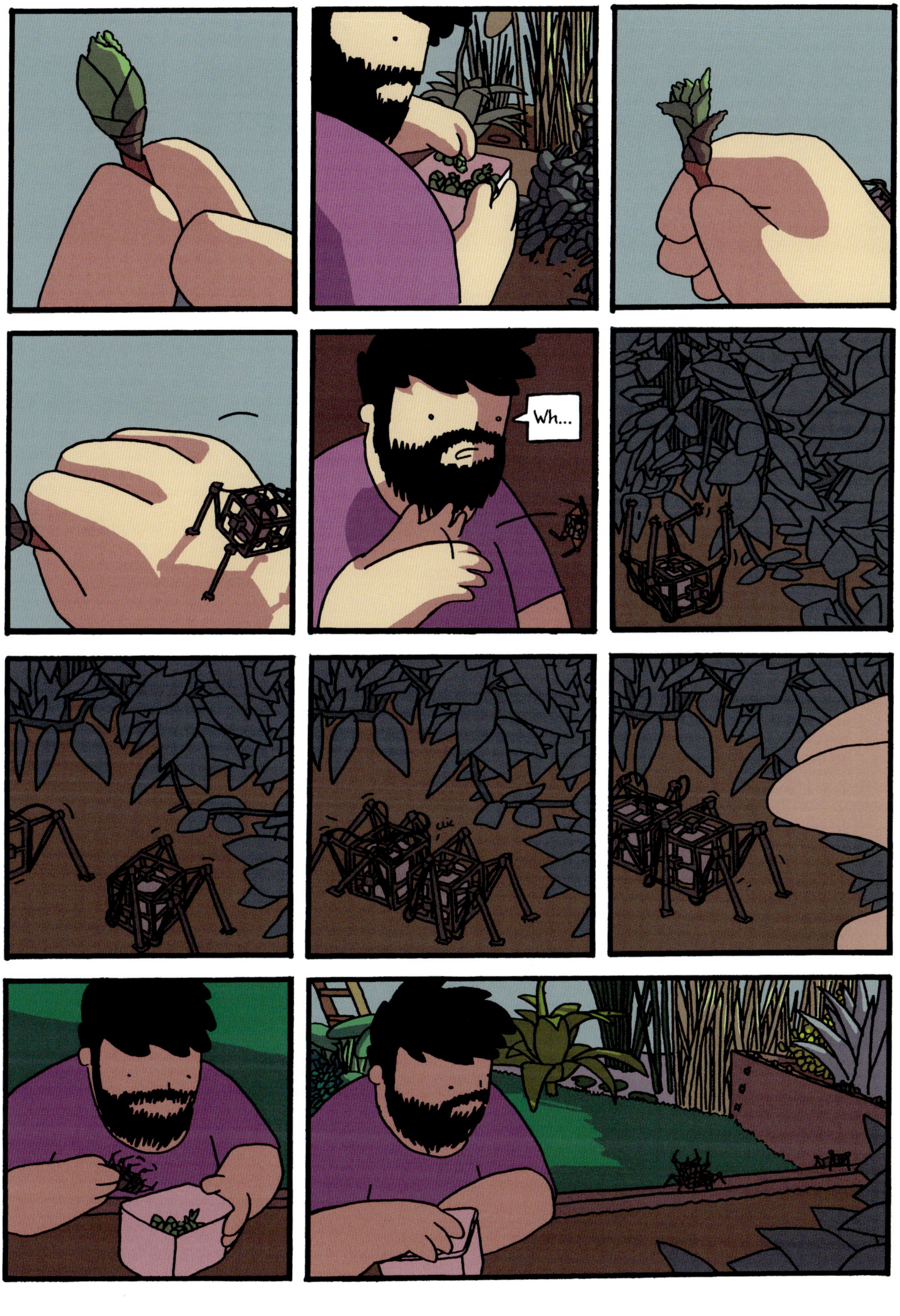
Wh...

Whoa!

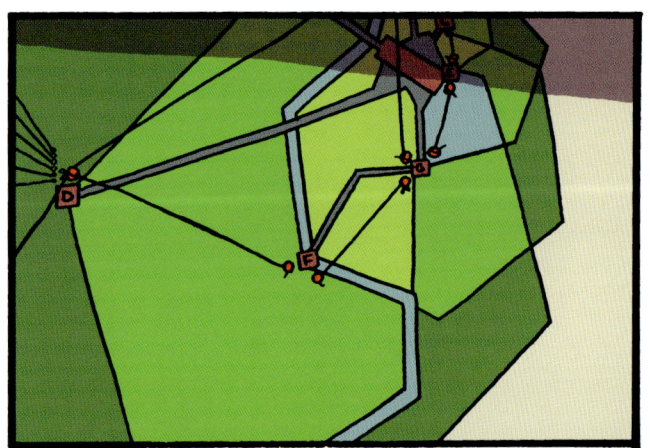

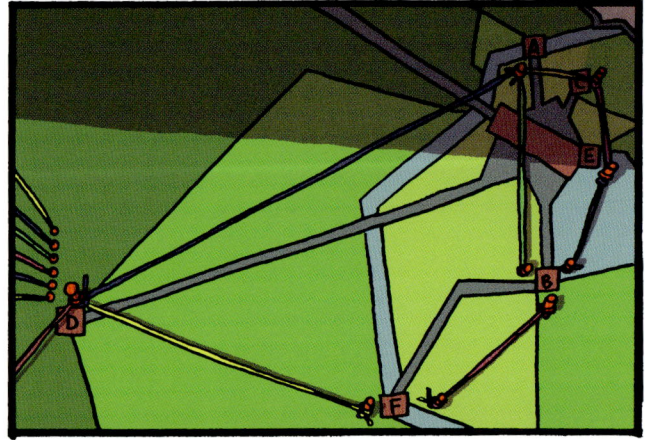

$$AF = \sqrt{(x_F - x_A)^2 + (y_F - y_A)^2}$$
$$FE = \sqrt{(x_E - x_F)^2 + (y_E - y_F)^2}$$
$$EC = \sqrt{(x_C - x_E)^2 + (y_C - y_E)^2}$$
$$CB = \sqrt{(x_B - x_C)^2 + (y_B - y_C)^2}$$
$$BD = \sqrt{(x_D - x_B)^2 + (y_D - y_B)^2}$$
$$DA = \sqrt{(x_A - x_D)^2 + (y_A - y_D)^2}$$

Algo - 23
While $L_{01} \neq \emptyset$, do
OK.
Yaawn!
Wha...?
BLOG:
USERNAME:
VINAIYPAN
PASSWORD:
N=1/P=0

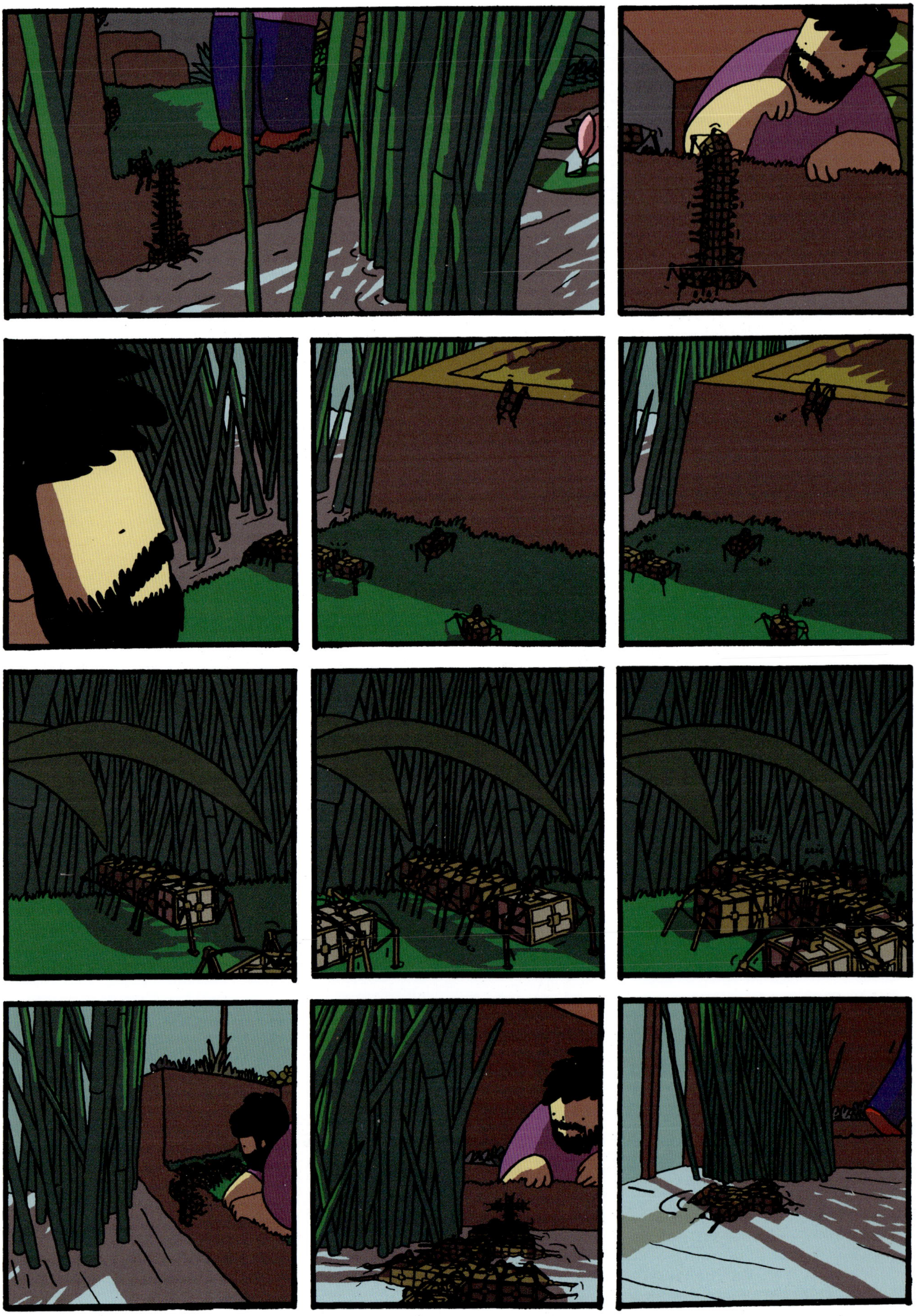

Oh, are those yours? Stéphane Douasy, Building F. Pleasure!
They went all the way to Building F? No way!
Hope they weren't bothering you.
Not at all. Microrobots that agile are a rare sight!
How did you make such a flexible structure?
I was inspired by origami. These are folded carbon fiber sheets.
Speeds up fabrication, too.

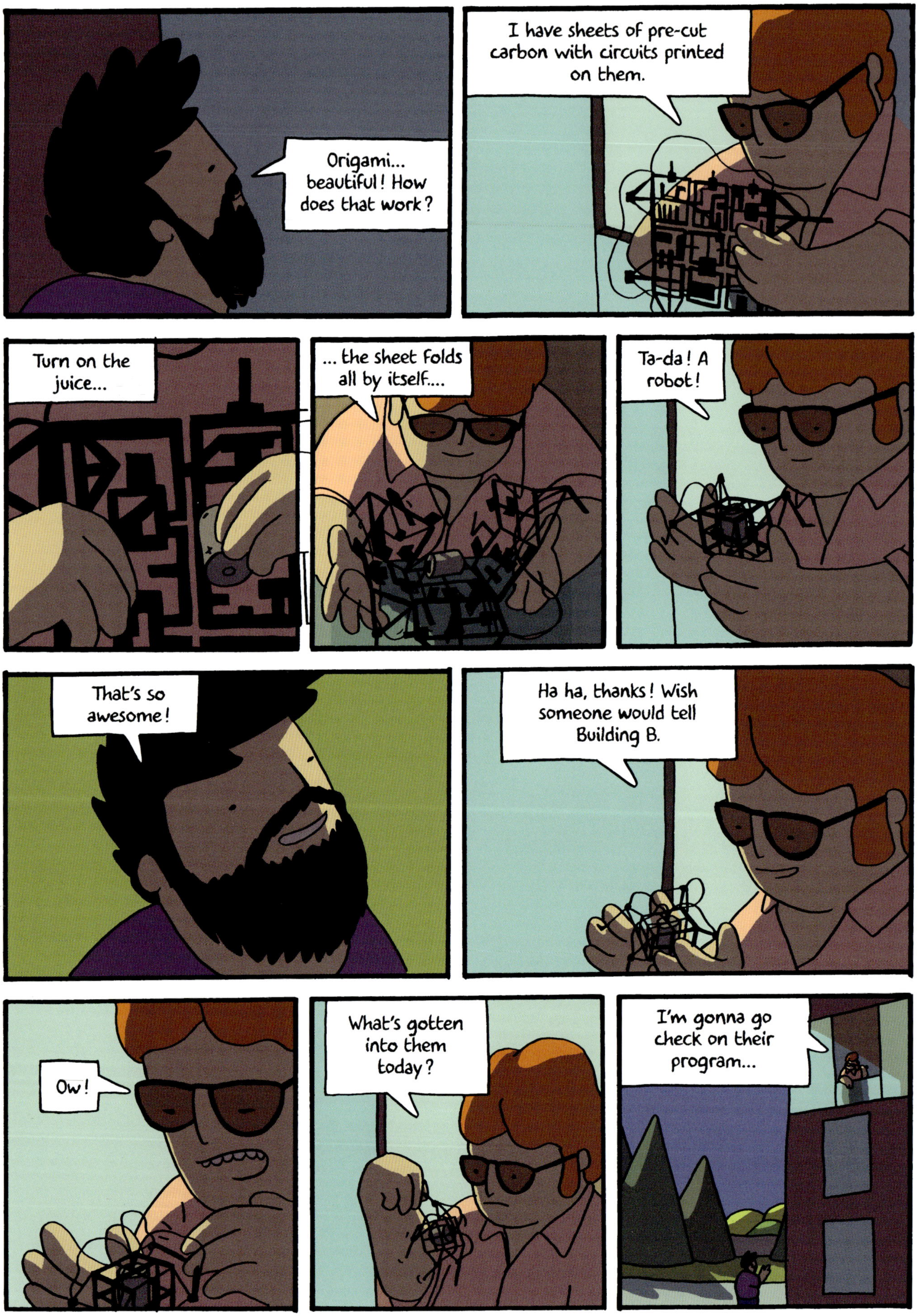
Origami... beautiful! How does that work?
I have sheets of pre-cut carbon with circuits printed on them.
Turn on the juice...
... the sheet folds all by itself....
Ta-da! A robot!
That's so awesome!
Ha ha, thanks! Wish someone would tell Building B.
Ow!
What's gotten into them today?
I'm gonna go check on their program...

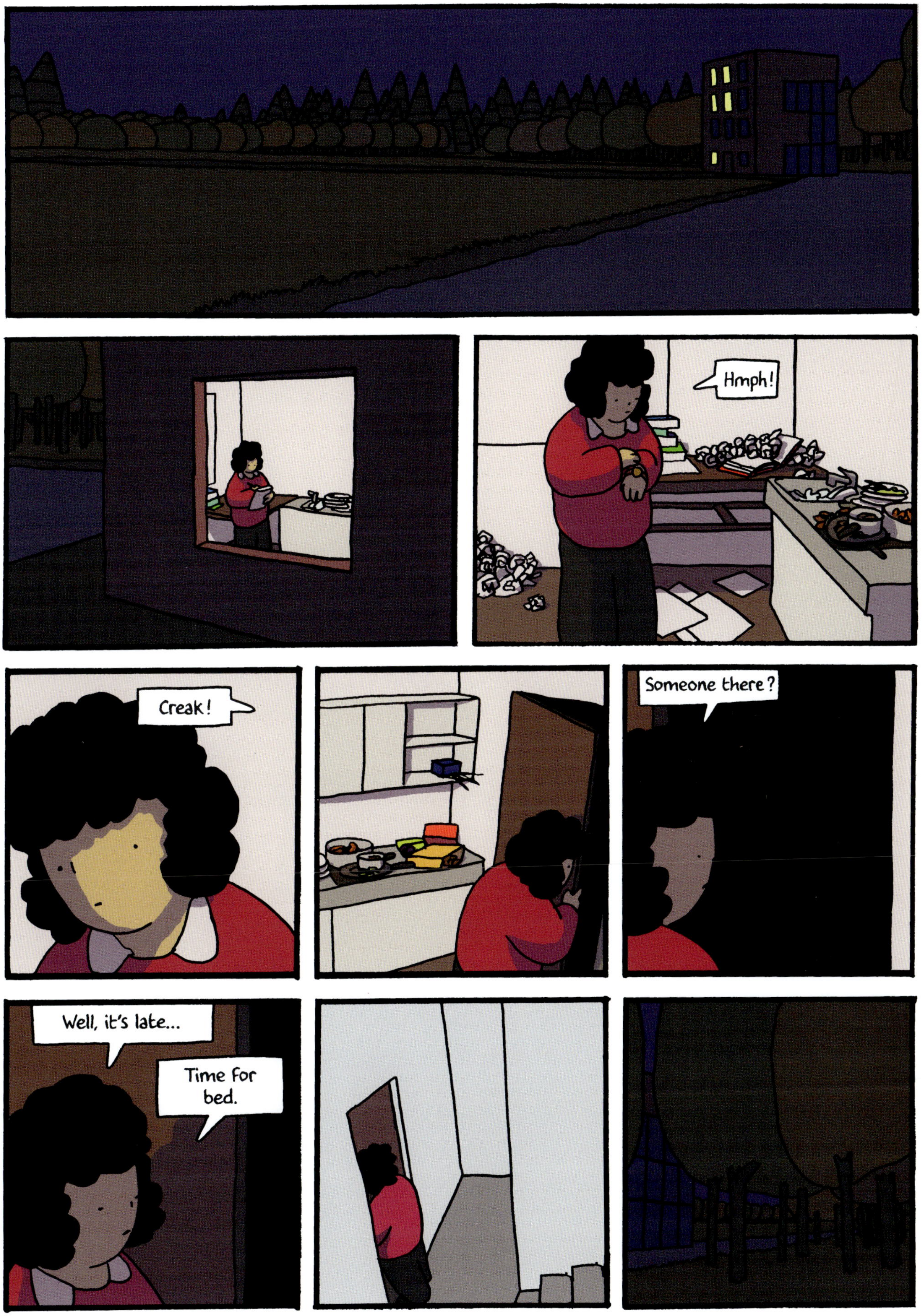
Hmph!
Creak!
Someone there?
Well, it's late...
Time for bed.

Boom! Boom!

Wham!

Knock knock.

Come in, Vilhem.

I got a sentence!

Really? Show me!

Stéphane, did you say anything to Han Fastolfe this morning?

7:53:12:
Pro. 16%
Diffraction of subject via interference from Stéphane Douasy and Han Fastolfe

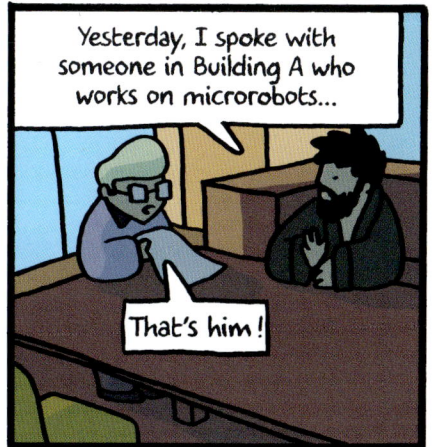

Uh... no. Haven't stepped out yet. / Sigh...	Who's this Fastolfe? / A roboticist in Building A.	Yesterday, I spoke with someone in Building A who works on microrobots... / That's him!
Hmm... It might be a coincidence.	Hold on, you think your program predicted that we were going to meet?	I don't know... / Plus, it's a day late.

What'd you discuss?

Uh, origami. / Nothing on that.	It gave me an idea, though... / For your spirals?	Oh, I've been done with those for a while. That was my dissertation. / Now I'm studying buds.

— So how do you explain spirals?

— Well, you'll have to read the article.

— Let's say it's not just a matter of evolution and natural selection...

— It's also a matter of spatial organization.

— The cells can't be arranged in any oth—

— Why that's it!

— There are far fewer possibilities... Some forms are inevitable.

— I have to revise my entire program.

— You're not so useless!

But I'm sure I turned the robots off when I left.
You must have forgotten.
How do you explain them organizing themselves and coming back from Building F?
Are you complaining about making advances?
I'm convinced someone snuck into my lab!
Why would anyone break into your lab and turn on robots?
I...
You'd do better to save your energy for research.
BAM BAM!
Help! The jellyfish are out of control. He's gonna drown!
I'll be there at once.
Hello? Dispatch a medical unit to Building D. It's an emergency.

Hurgl!
How is he?
He was unconscious for a while...
May I speak to you in private?
You know that's out of the question!
You will see to him personally. The Institute has all the necessary equipment!

Tell me what happened, Chizuru.	He fell in the water and... ...the robots attacked him.	It was like an actual immune system response.
Perfect! That's what you were going for, right?	Yes, but... Back to the lab! Try to figure out why it worked.	We'll take care of him.

BZZ!

P = NP?
ANSWER: NO
P = NP
WHY P IS NOT EQUAL TO NP.
WHAT IF P = NP?
P VS NP
Sigh
This is beyond me...
BLOG:
USERNAME:
VINAIYPAN
P VS NP
Vinaiy Paniandy's blog
Sign in
Username : Vinaiypan
Password : *******
P VS NP
Vinaiy Paniandy's blog
You are now signed in
New page
New post
Enter forum

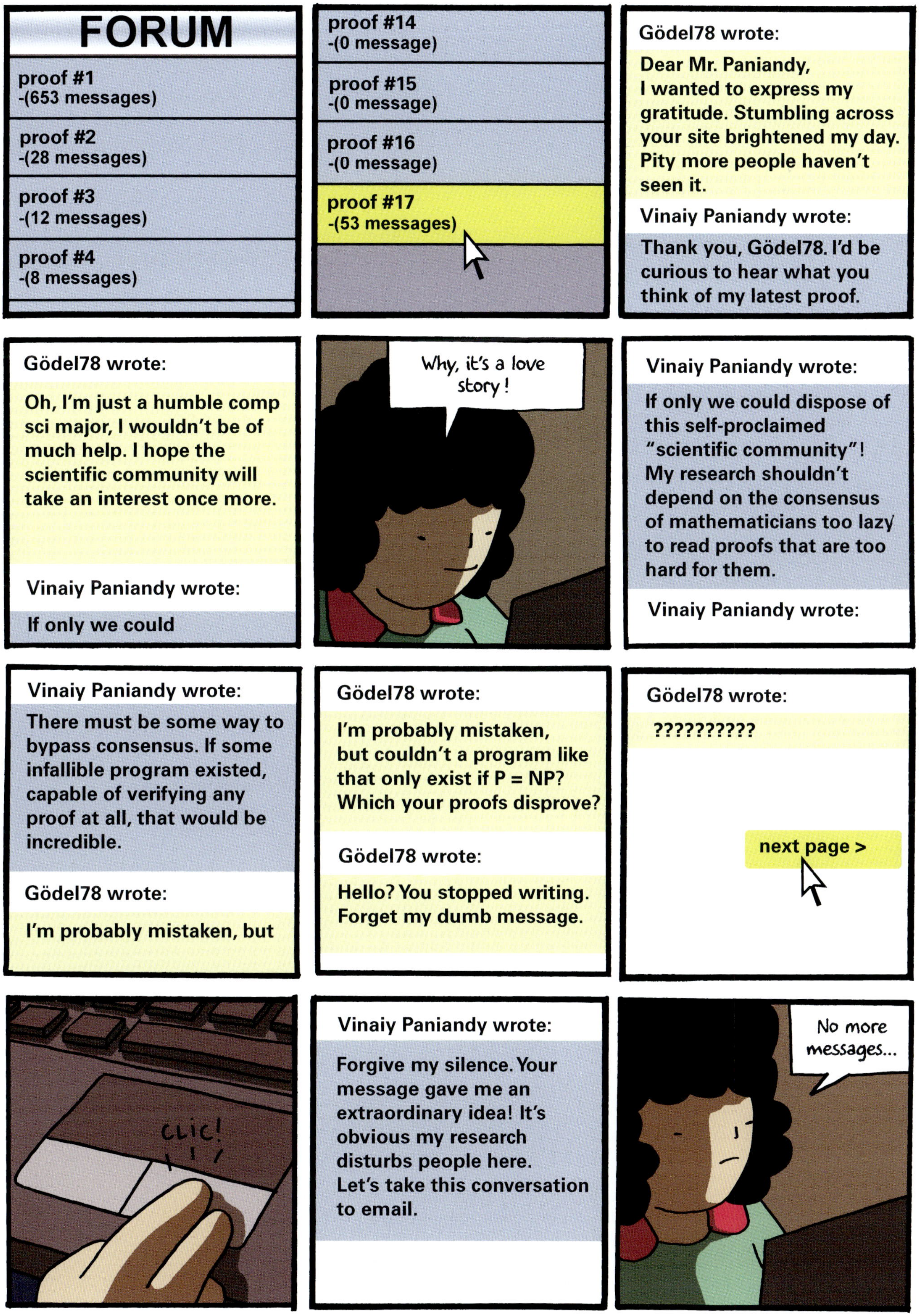

FORUM

proof #1
-(653 messages)

proof #2
-(28 messages)

proof #3
-(12 messages)

proof #4
-(8 messages)

proof #14
-(0 message)

proof #15
-(0 messages)

proof #16
-(0 messages)

proof #17
-(53 messages)

Gödel78 wrote:

Dear Mr. Paniandy, I wanted to express my gratitude. Stumbling across your site brightened my day. Pity more people haven't seen it.

Vinaiy Paniandy wrote:

Thank you, Gödel78. I'd be curious to hear what you think of my latest proof.

Gödel78 wrote:

Oh, I'm just a humble comp sci major, I wouldn't be of much help. I hope the scientific community will take an interest once more.

Vinaiy Paniandy wrote:

If only we could

"Why, it's a love story!"

Vinaiy Paniandy wrote:

If only we could dispose of this self-proclaimed "scientific community"! My research shouldn't depend on the consensus of mathematicians too lazy to read proofs that are too hard for them.

Vinaiy Paniandy wrote:

Vinaiy Paniandy wrote:

There must be some way to bypass consensus. If some infallible program existed, capable of verifying any proof at all, that would be incredible.

Gödel78 wrote:

I'm probably mistaken, but

Gödel78 wrote:

I'm probably mistaken, but couldn't a program like that only exist if P = NP? Which your proofs disprove?

Gödel78 wrote:

Hello? You stopped writing. Forget my dumb message.

Gödel78 wrote:

??????????

next page >

CLIC!

Vinaiy Paniandy wrote:

Forgive my silence. Your message gave me an extraordinary idea! It's obvious my research disturbs people here. Let's take this conversation to email.

"No more messages..."

71

"Simon?"

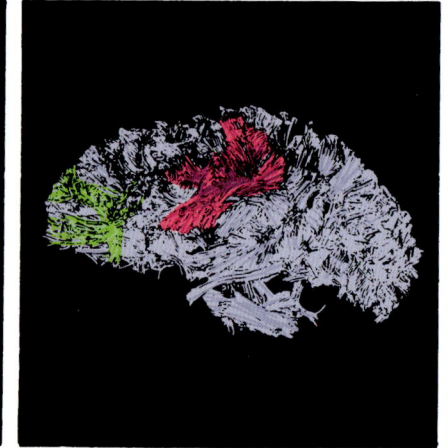

"I've got a program that might interest you..."

TZZZ!
07:12:12
Prob.: 75%
Subject asleep.
07:12:12
Prob.: 75%
Subject asleep.
07:12:13
Prob.: 81%
Subject awake.
Hmm... Stats are good, but still too slow!
09h12m 06s64c
TZZZ!
end simulation
08:07:53
Prob.: 69%
Subject visited by Louise François and Stéphane Douasy.
08:08:00
Prob.: 50%
Death of subject.
09h32m 03s03c
KLAK!

Stéphane?

Ha! Stéphane, is Louise with you?

No. What's up?
Were you two going to come see me?
Is this your program again?
Look at this!
See that probability rate?
08:07:53
Prob. : 69%
Subject visited by Louise François and Stéphane Douasy.
08:08:00
Prob. : 50%
Death of subject.
Well, it makes sense, for once...
But I wasn't planning to come see you. I'm taking photos.
We can go ask Louise if you want.
Oh! It's open...
Louise?
No one's home.
Come quick! I'm downstairs!

Panel	
"It's all gone!" / "What?"	"His lab! It's empty!" / "Something fishy's going on!"

"He must've left the Institute..."

"I checked out his blog."

"He posted 17 proofs in a row, and then three months ago... zilch."

"His last message implies he feared for his research."

"Feared what?"

"No idea. But you gotta admit it's fishy."

"17 publications in 6 months?! That's obsessive!"

"On his blog, he said his work disturbed people."

"Let's not get paranoid."

"And when I went to ask Martin..."

"... all his stuff vanishes like he never existed!"

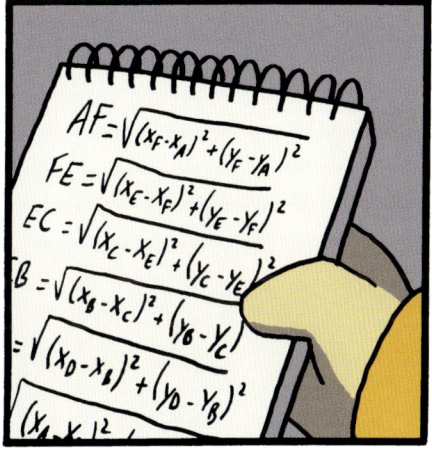

I think he proved P = NP!	Pshaw.	I thought I explained why that problem was beyond most mortals' reach.
Look at what I had time to jot down before it all disappeared.	$AF = \sqrt{(x_F-x_A)^2 + (y_F-y_A)^2}$ $FE = \sqrt{(x_E-x_F)^2 + (y_E-y_F)^2}$ $EC = \sqrt{(x_C-x_E)^2 + (y_C-y_E)^2}$ $B = \sqrt{(x_B-x_C)^2 + (y_B-y_C)^2}$ $= \sqrt{(x_D-x_B)^2 + (y_D-y_B)^2}$	Hah! He picked the traveling salesperson as an NP-complete problem.
Uh, translation, please?	The traveling salesperson problem is: "What is the shortest path connecting a series of points in space?"	Here, he's working from a map of the Institute.
If he finds an algorithm for figuring out the shortest path on the first try...	...then he'll have proved P = NP.	Oh, OK.

Weird. In his articles, Paniandy was trying to prove there was no algorithm for that problem. Other scientists refuted his proofs each time.	But I wrote down something that looks like an algorithm. Show me!

Here.	Why would his discovery "disturb" certain people?	If that algorithm works, it could be worth millions! It'd literally explode the possibilities of computing.
He uses fractals?	Whoa... I don't get it, but it sounds terrific!	If this works, it'd let me speed up my program considerably!

Yes, and revolutionize mathematics!	Could I make a copy? I'll try and adapt it...

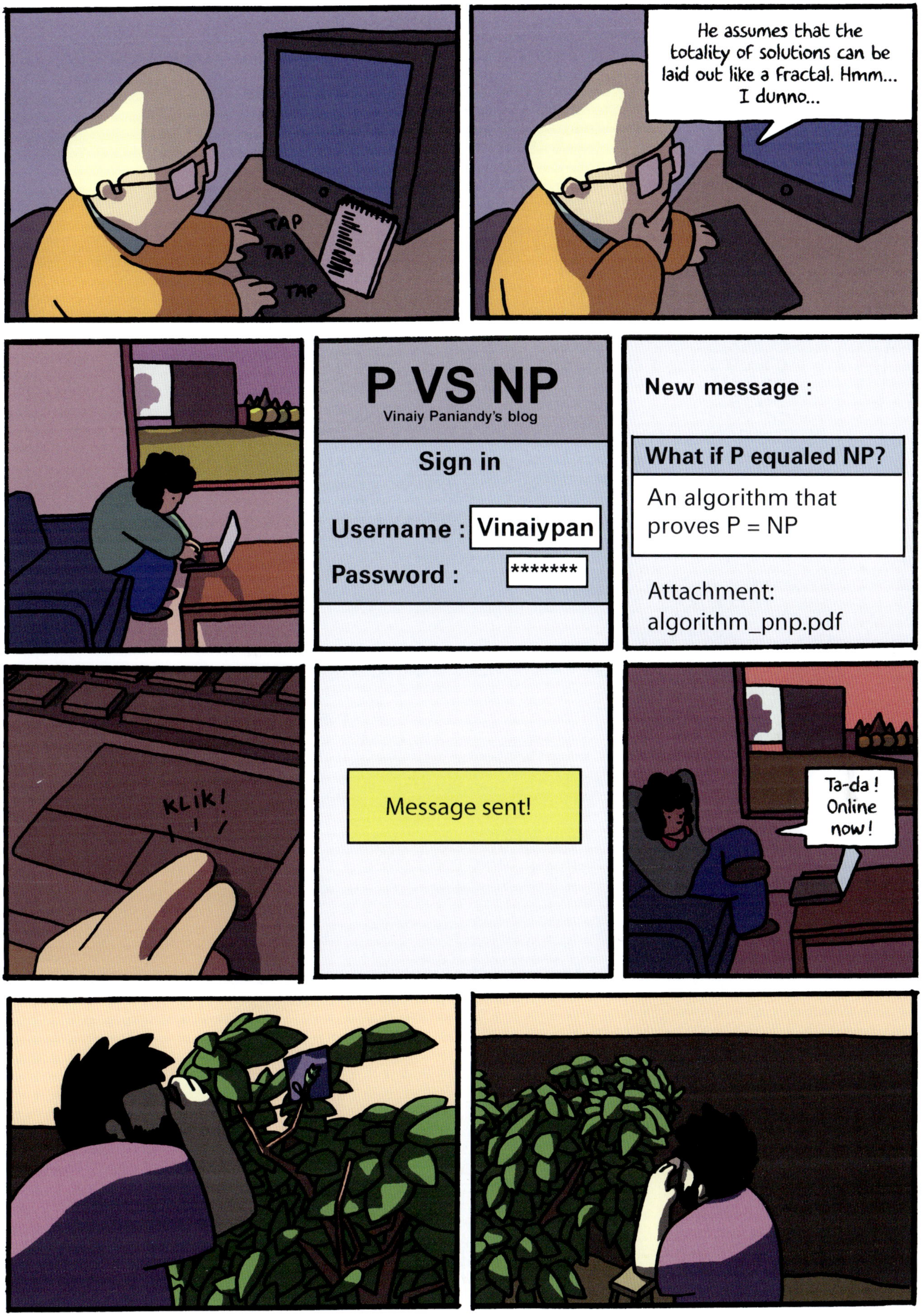

TAP
TAP
TAP
He assumes that the totality of solutions can be laid out like a fractal. Hmm... I dunno...
P VS NP
Vinaiy Paniandy's blog
Sign in
Username : Vinaiypan
Password : *******
New message :
What if P equaled NP?
An algorithm that proves P = NP
Attachment: algorithm_pnp.pdf
KLIK!
Message sent!
Ta-da! Online now!

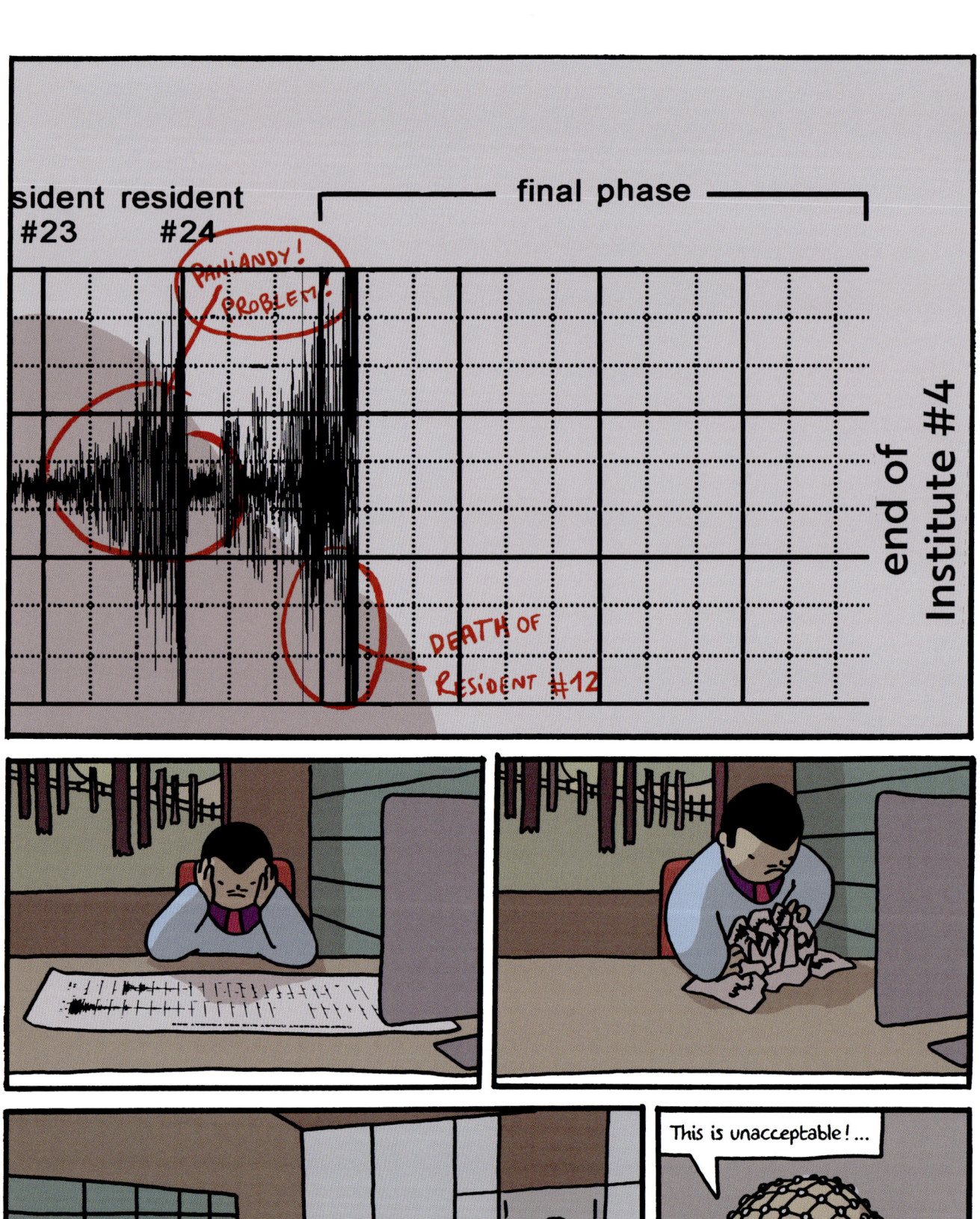

resident #23 resident #24 — final phase —

PANIANDY! PROBLEM?

DEATH OF RESIDENT #12

end of Institute #4

"This is unacceptable!..."

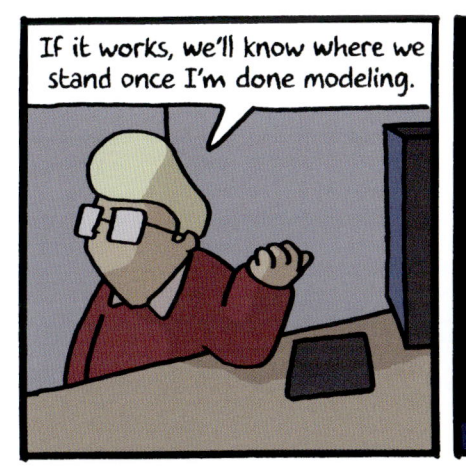

Well... never hurts to try.

Bam bam!

TAP!

I posted his article...

...so the community can assess the algorithm.

If it works, we'll know where we stand once I'm done modeling.

How about coming with me to see Martin? Two of us might have more leverage.

OK.

Martin?

My God!

"My God, it's... Crap, what's his name again?" / "Simon Burrick, neurosciences."	"You think he's dead?"	
"I don't know. Check his pulse."		"What happened here?"
"Wha—?"	"We have no idea! We just got here!"	"Where's Martin? He called for me—"
"We haven't a clue! We just—" / "Oh no!"	"He's dead."	

What happened to him?	What are you doing?	Trying to find what killed him.
Simon recorded all his brain activity 24/7.	There must be some trace of his death. Booting...	The system has just recovered from a serious attack. Please start antivirus protection. !
His computer's been hacked!	Apparently that caused an aneurysm.	What, you mean his laptop killed him? Seems so...
Was it... murder?		Look, I think we're in danger.

"That's the second death today."

"What?"

"Victor Blumberg drowned earlier..."

"What are you doing?"

"Looking for clues."

"Don't bother. Chaos is rearing its ugly head."

"What do you mean?"

"When Martin and I were the only two here, he told me the Institute was designed to descend into chaos."

"The selection of researchers, their order of arrival, their lab assignments..."

"All of it engineered to create a simulation, an instability conducive to major scientific discoveries."

"Chaos theory applied to a group of human beings. Gross."

"Martin's job was to limit the chaos to reasonable proportions."

"Look, I know I have no proof, but it all leads me to believe Martin's behind this mess."

"Martin, a murderer? One thing's for sure: we're all at risk."

"And Martin's disappeared? We have only one choice."

"What?"

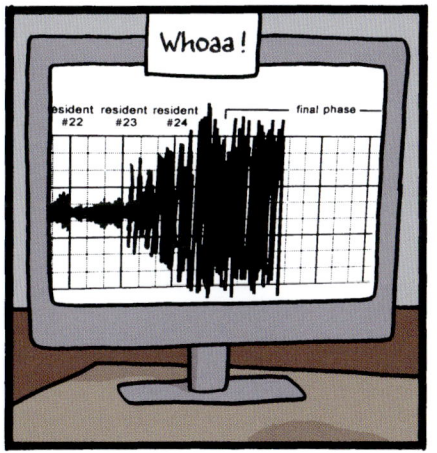

Panel 1: "Call in the clean-up crew!" / "Are you nuts? If we do that, the Institute's over! All our research..." / "She's right!"

Panel 2: "I don't feel like dying." BEEP!

Panel 3: "Hello?" / "Hello. Please select the reason for your call." / Automated system.

Panel 4: "To schedule a visit from the clean-up crew, press 8." BEEP!

Panel 5: "Please start the Institute's chaotic behavior monitoring program." / "Uh..."

Panel 6: Analyzing chaos... Klik!

Panel 7: Whoaa! / resident #22 resident #23 resident #24 final phase

Panel 8: "Now click on Clean-Up." / **Clean-Up** (this operation is irreversible)

Panel 9: Clean-Up in 35:59:57 / "Your records indicated that clean-up will be necessary in 36 hours. Until then, we request that you remain in your laboratories with the door locked. Do not offer any resistance when our agents arrive. Goodbye. Klik!"

Panel 10: OOOWEEEO OOOOWEEE EEOOOWEE

"Where's Vilhem?"

"He just left."

"Well... now we wait."

OOOWHEEEOOOOOWHEEEEEOOOWHEEEOOOOOWHEEEEEOO

OWHEEEOOOOOWHEEEEEOO

EEEE... In compliance with Article 224-B of your Institute residency contract...

... a state of emergency has been declared. Our clean-up crew will intervene in 36 hours...

Until then, we request that you remain in your respective laboratories with the door locked...

and avoid all contact with other researchers. We advise you to prepare for departure...

Any sign of resistance when our agents arrive will be seen as a refusal to comply...

We remind you that our agents are authorized to use force if necessary...

OOOWHEEEOOOOOWHEEEEEOOOWHEEEOOOOOWHEEEEE In compliance with Article 224-B of your Institute residency contract, a state of emergency has been declared. Our clean-up crew will...

intervene in 36 hours. Until then, we request that you...

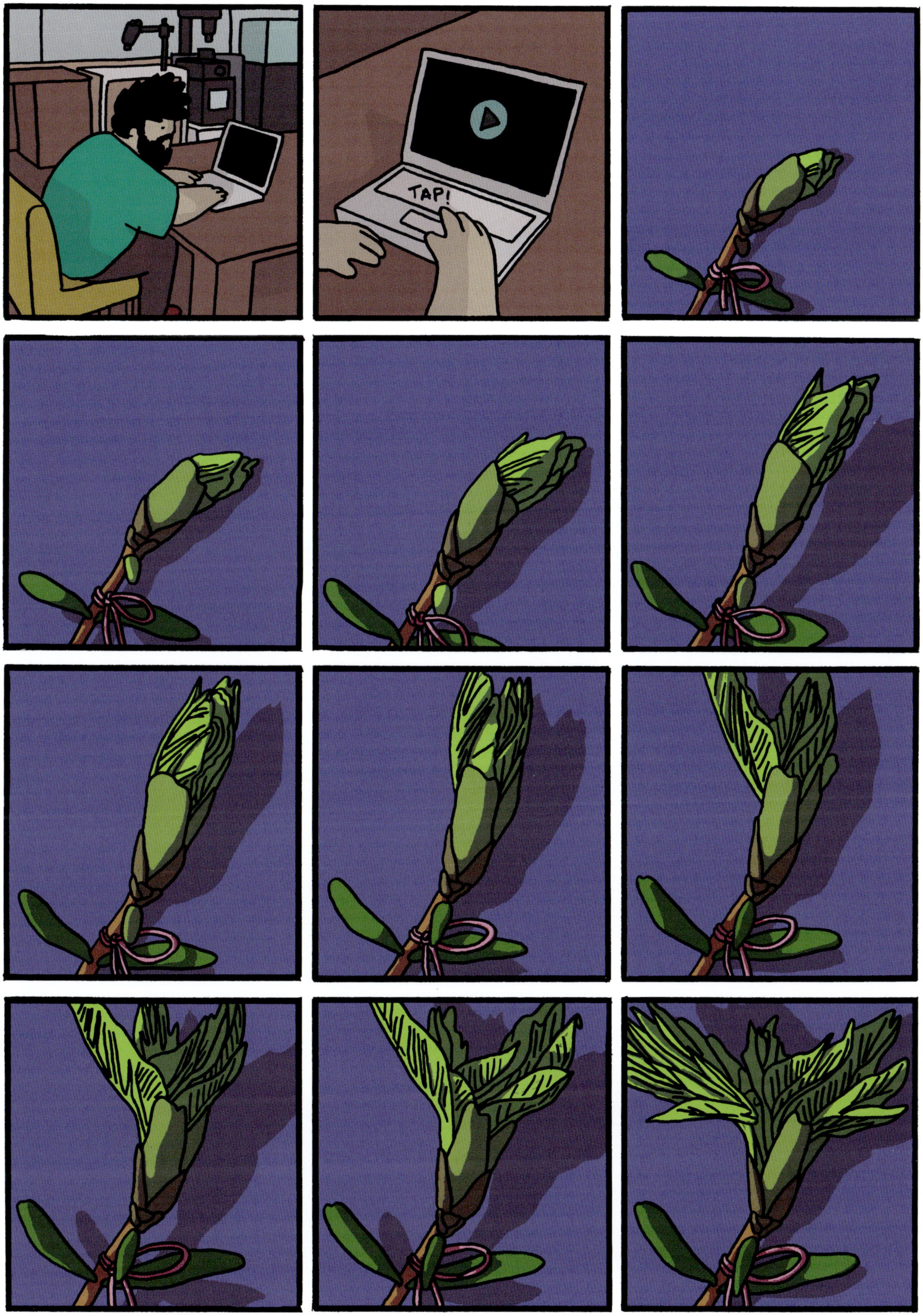
TAP!

OOOWEEEEOOOOWEEEO OWEEEEOOOOWEOO	OOO WEEEO OWEEEE EEEOO OO — Those morons!

Paniandy!

Paniandy!

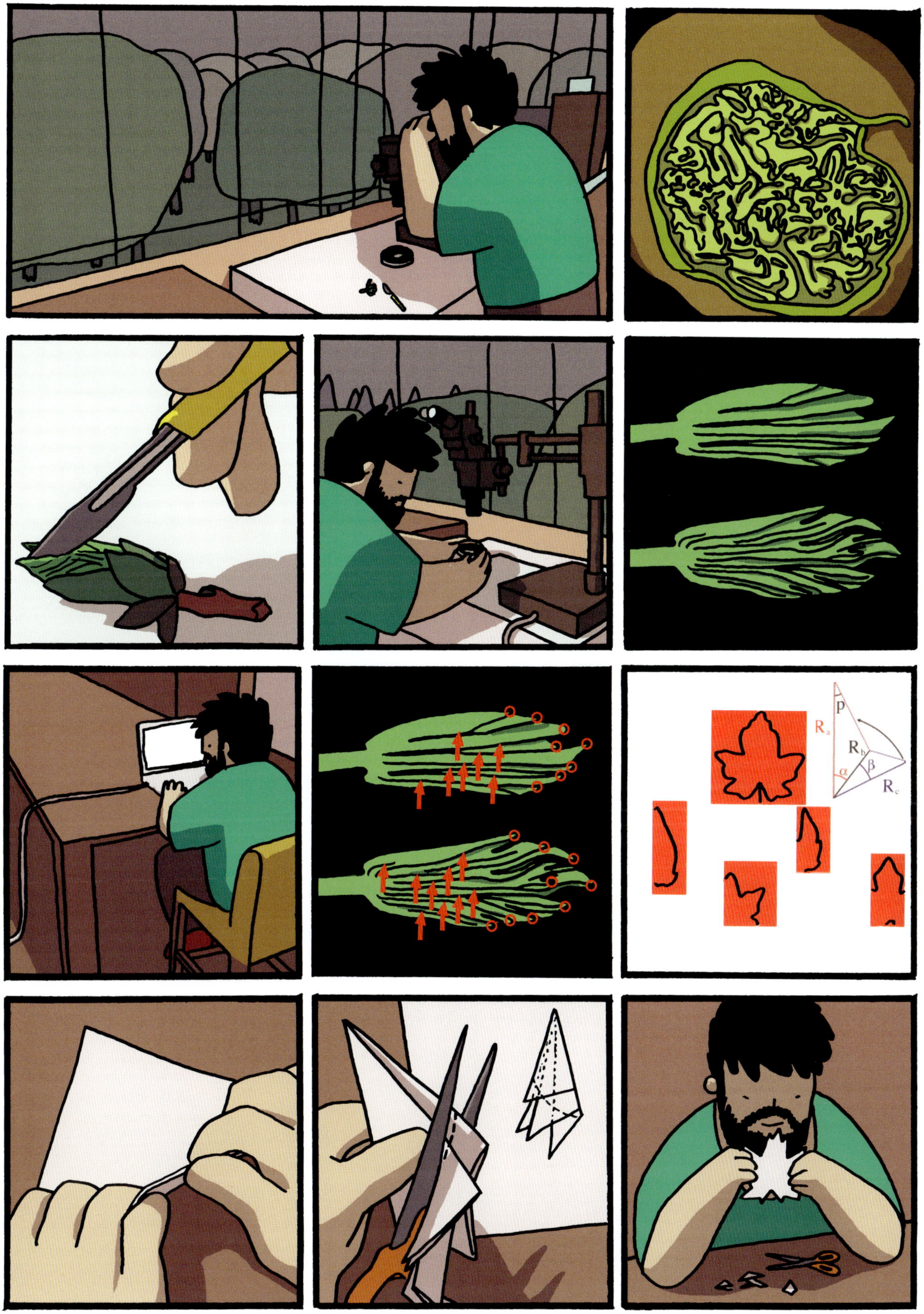
p
R_a
R_b
α
β
R_c

WHAM!
Those morons!
They've sounded the Institute's death knell.
Drop everything. Save what can be saved and pack your bags.
But... my plants?
No idea... No one knows what they do with our research once an Institute ends...
Shit.
You said it.
Whatever you do, stay inside!
Be a good boy, wait for them to knock on your door, and do whatever they say!
See you in two days.

"What'll you do with the body?"

"Put it in cold storage with the other one. Once the Institute shuts down, they'll investigate."

"So why do you think Martin did this?"

"Did you know one of the four residents in our building vanished?"

"Really?"

"Well, actually... we just never saw him."

"I made the mistake of asking Martin about him, but..."

"... he wouldn't give me any info at all."

"I poked around a bit, and that's when things got complicated..."

"I found out Paniandy, the resident in question, definitely solved one of the seven Millennium Prize problems."

"Work that challenges the certainties of most mathematicians..."

"His findings prove that algorithms can answer many more questions than believed."

"We could create a truly advanced AI."

"I just know Martin got rid of him..."

"Either to steal his research, or because it was too disturbing."

"Well, if it's any consolation, Paniandy paid me a visit just two days ago."

"What?"

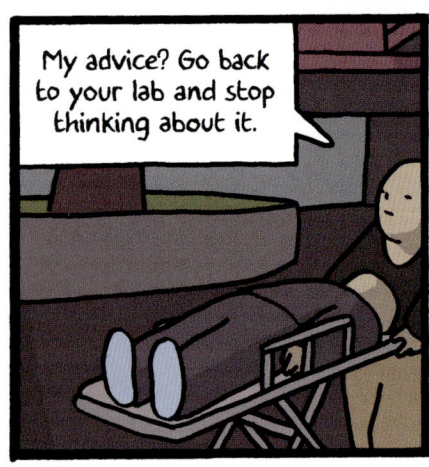

"You've met him?"

"He's alive?"

"He smelled bad and looked unstable, but definitely alive."

"I gave him some antidepressants, and he vanished again."

"Oh..."

"Fear not, the clean-up crew will take care of all this."

"Yeah..."

"Where could he have been hiding all this time?"

"Maybe Martin had him locked up?"

"My advice? Go back to your lab and stop thinking about it."

"Stay safe from the chaos, and see you in..."

"35 hours."

"So... what does he look like?"

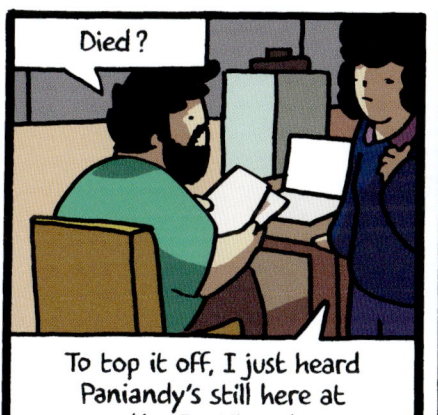

Paniandy? Can't miss him. Huge beard, curly hair, big round glasses.

Not too pissed off?

Nah, I'm OK. My research was only starting out anyway. It's all in my computer.

Sorry, but two people have died, and Martin's AWOL...

Died?

To top it off, I just heard Paniandy's still here at the Institute!

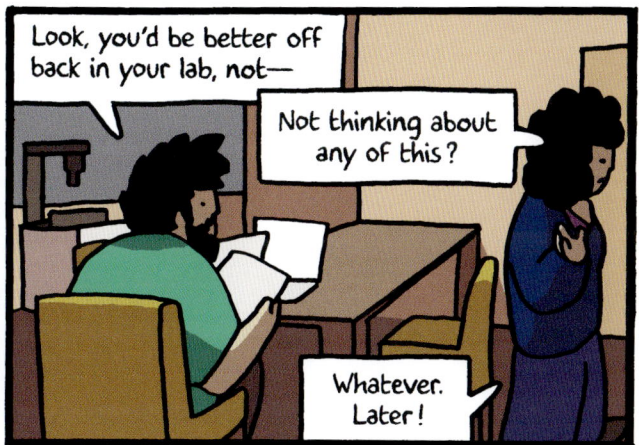

Enough with Paniandy! You're paranoid!

So how'd they die? An accident?

I'm the only person here who finds two deaths and two disappearances in just a few days weird! That's why I'm paranoid!

Look, you'd be better off back in your lab, not—

Not thinking about any of this?

Whatever. Later!

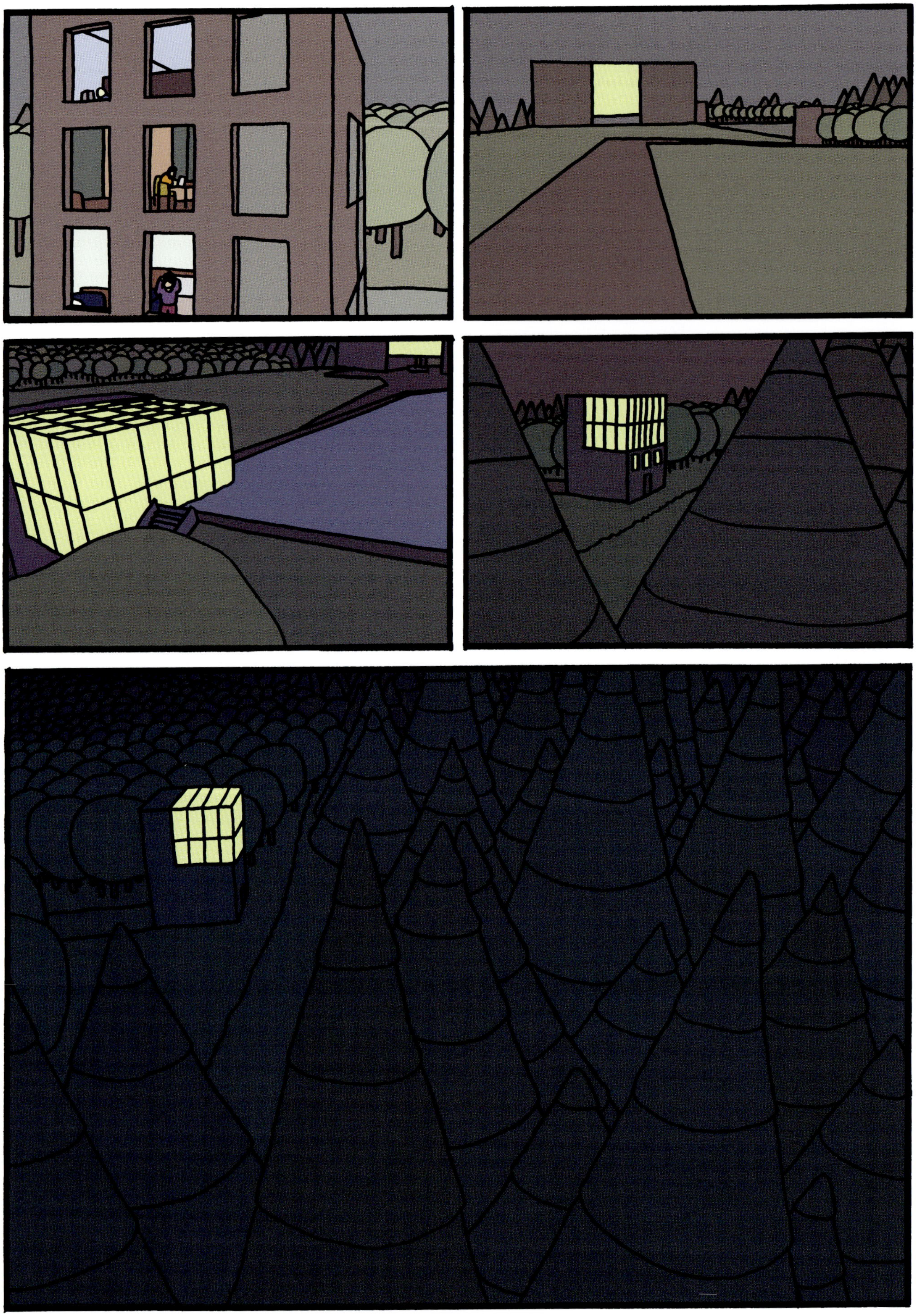

TZZ!
07:23:47 prob. 92%
Subject reads printout
07h23m 47s00c
07:23:48 prob. 94%
Subject checks watch.
07:23:50s prob. 90%
Subject reads printout again.
07:23:53 prob. 91%
Subject's heart rate speeds up.
07:24:00 prob. 89%
41% increase in adrenaline and cortisol levels.
07:24:08 prob. 95%
Subject exits laboratory.
07:24:19 prob. 96%
Subject exits experiment's effective area.
21:07:13 prob. 98%
Subject interacts with Vinaiy Paniandy.
21:21:12 prob. 100%
Death of subject.

P VS NP : Vinaiy Paniandy's

Your posts (18)

Your posts (18)

- "What if P equaled NP?"

comments : 1

Gödel78 wrote:

Dear Vinaiy, I'm so glad to see you've emerged from your forests at last :)
So this is it! You've done it! An algorithm for an NP-complete problem! And your other program?

Him again...

It's like we're the only two people in the world who care about Paniandy's work!

CLAP

Where are you hiding, Paniandy?

Hey!

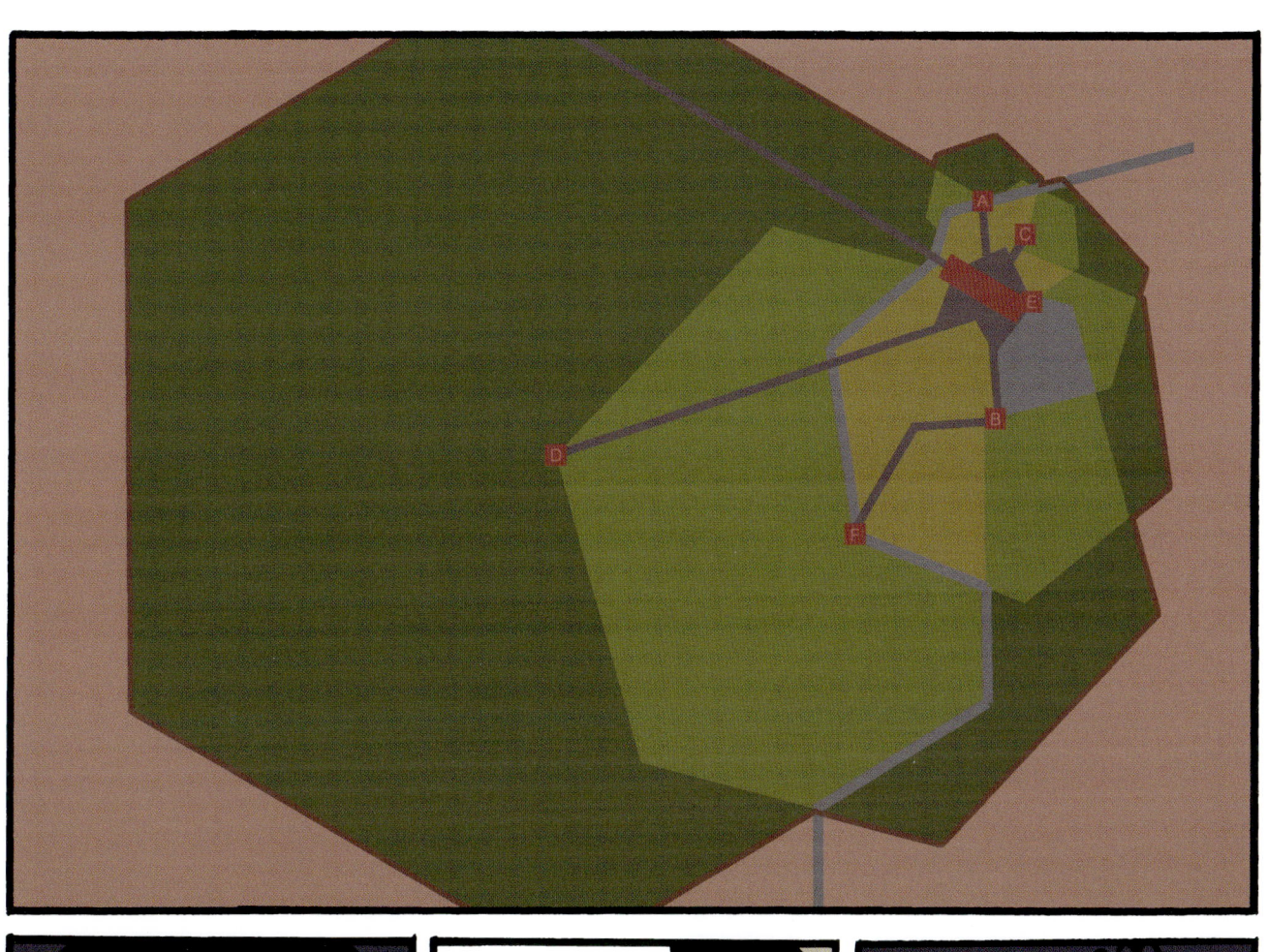

"Emerged from your forest"!

He's in the pine woods!

TAP TAP

— Stéphane?

— Louise? They said not to talk to each other!

— I know where Paniandy is.

— I missed an obvious clue!

— This location in the forest doesn't correspond to any Institute building.

— Feel up to a field mission?

— Are you kidding? Looked outside recently? I'm staying right here!

— What do you mean?

— That's just the tip of the chaos you were talking about.

Whatever. I'll ask Vilhem.

But...

Go back to your lab, Louise!

Vil...

He's gone!

15:17:22 prob. 89%
Subject heads deep into the woods.

15:28:10 prob. 92%
Martin Sorokin enters subject's field of vision.

Hey, his thing's working!

I just know Martin's in on it!

21:07:13 prob. 98%
Subject interacts with Vinaiy Paniandy.

21:21:12 prob 100%
Death of subject.

Crap! That's in an hour!

21:21:12   prob 100%
Death of subject.

Things are starting to deteriorate.

AAHH!

Aahh!

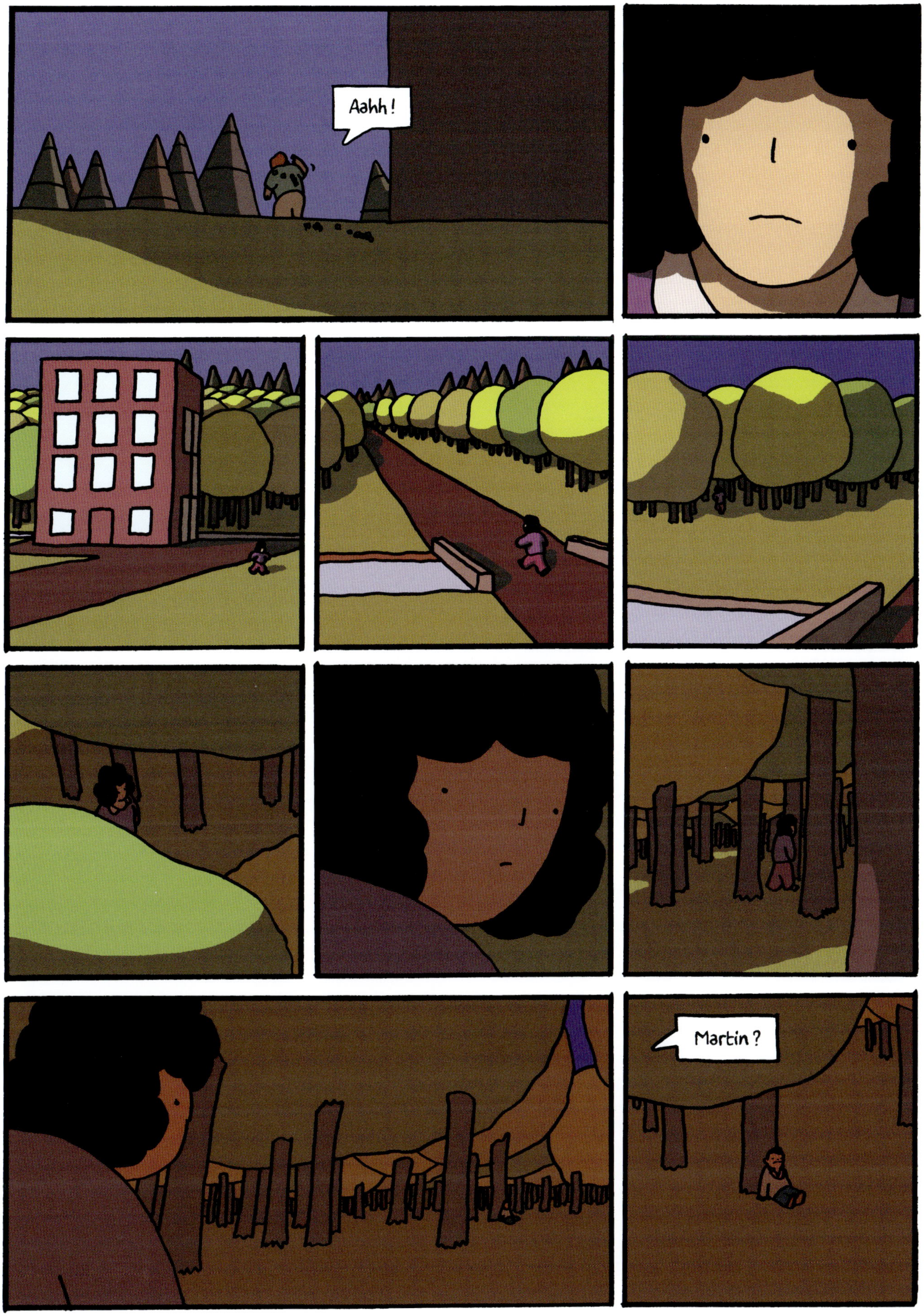
Aahh !
Martin ?

Back to your lab, Louise!	Enough lies. I want some answers.	What are you scheming? What's going on with Paniandy? And Vilhem?
Ha ha! You should revise your hypothesis.	My only scheme was to protect the Institute from Paniandy.	He's out of control. Chaos skyrocketed after he showed up!
But...	His so-called discovery caused the deaths of two residents!	Hngh.
A— are you hurt?!	It was Vilhem. We fought. I don't know what ideas you put in his head. I've never seen him like that.	Vilhem?

"I wouldn't tell him where Paniandy was."	"He didn't appreciate that..."	"Well, duh. Look at his program. It's working!" "That's absurd!" "Wait..."
"It predicted the two of you would meet!"	"OK, fine. So what if it did?"	"It also says Vilhem will die in under an hour, just after meeting Paniandy!"
"Don't tell me you believe in that nonsense. There's no predicting the future."	"But if anyone believes in it, it's Vilhem! I have to find them at once!"	"Louise!" "Yes?" "Be careful. Vilhem... has my gun..." "Ok."
"Oh, Martin? I was lying. My research was never going anywhere."	"I knew it."	

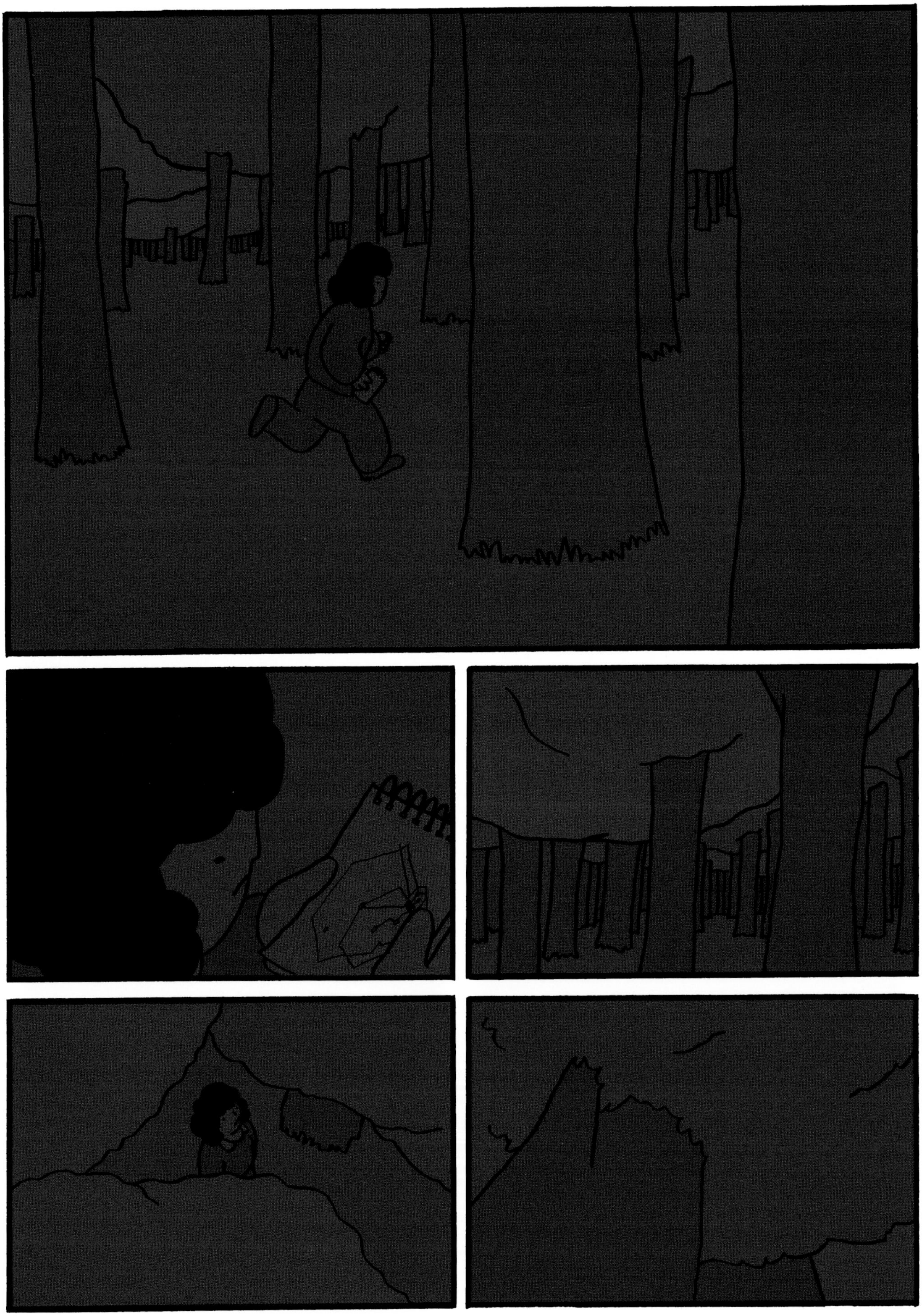

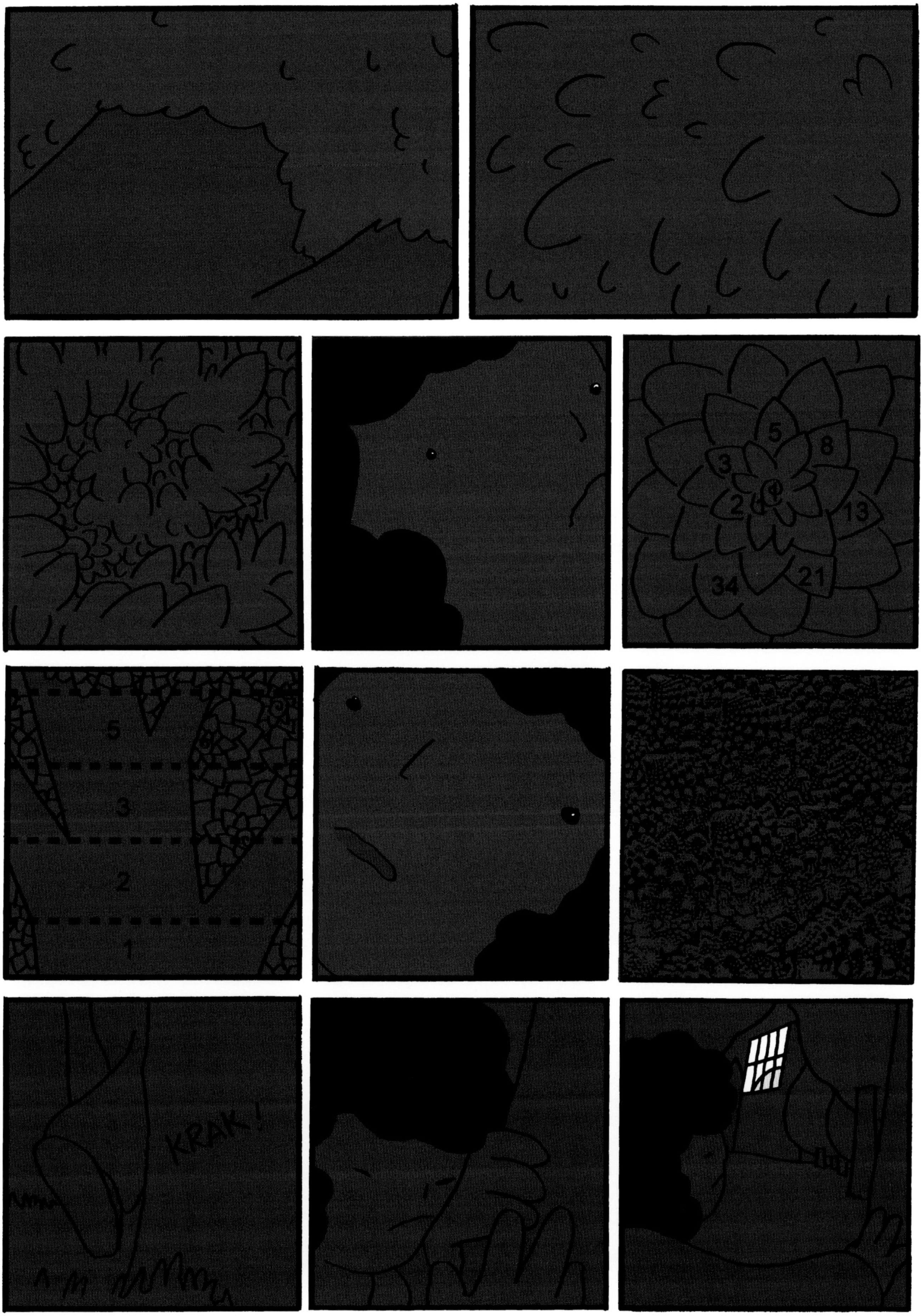
5
8
3
2
13
21
34
5
3
2
1
KRAK!

	Hello.	Uh... are you the astrophysicists?
		Are you feeling OK? You look all pale.
Oh, I'm OK. You wouldn't have any binoculars, would you?	I'll go find a pair. — You shouldn't be outside... — We shouldn't even talk to her...	

Here.
Thanks a lot.
See anyone coming or going from the woods?
Yes. Martin and another resident are often tinkering nearby...
Beard, curly hair, and glasses?
That's right. But I don't know if going out now is a good idea.
I'm a big girl, thanks. Where'd you see them?
It's easy. Let me explain.
... and then you'll reach a clearing with some fallen trees. There.
Here's a flashlight.

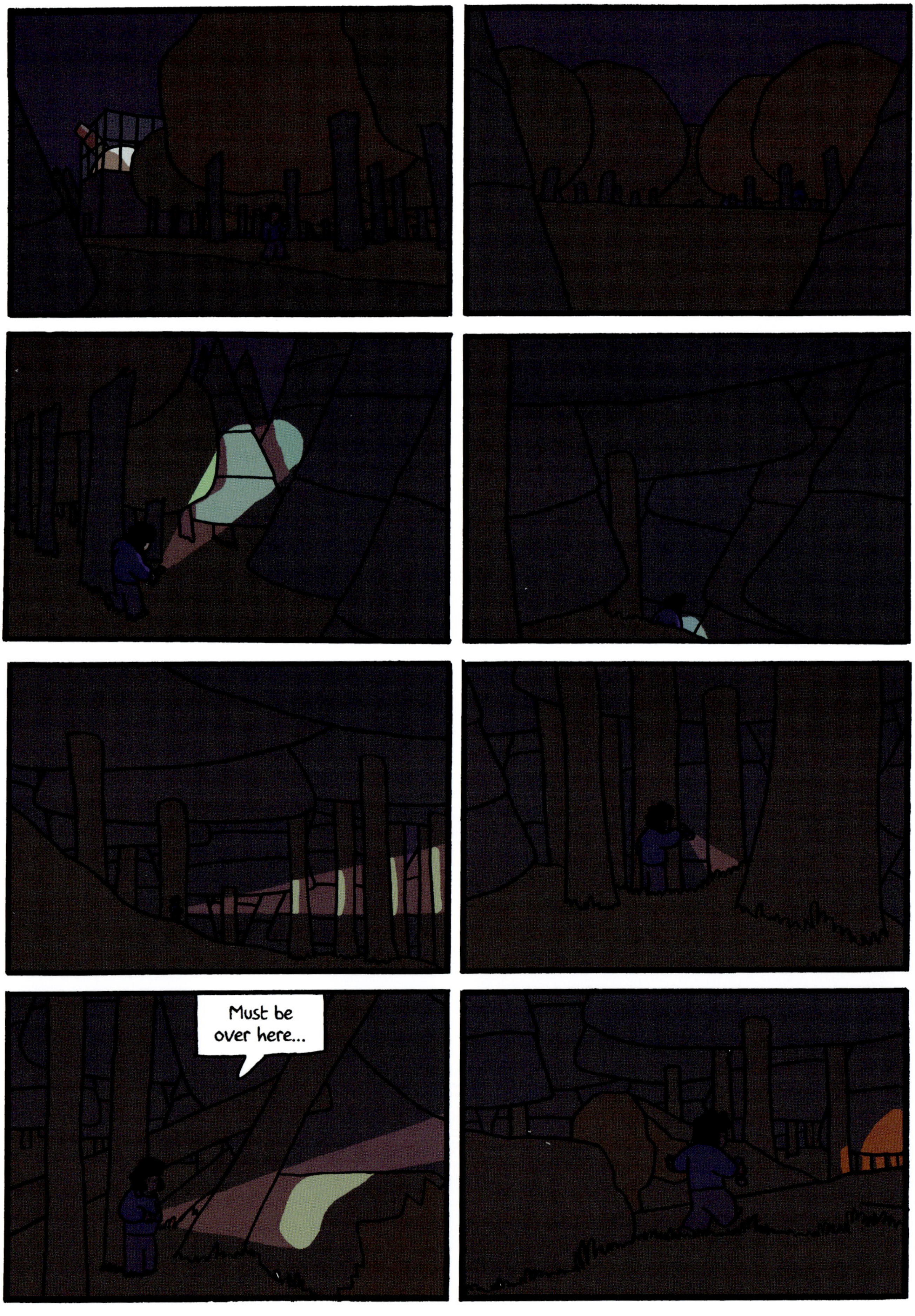
Must be over here...

So this is where you've been hiding this whole time!

I haven't been hiding. I was told I had to distance myself.

The community doesn't like my research.

Your algorithm, you mean?

Yes. The one you stole!

I never stole a thing! I don't even get what it's for.

What it's for? Ha ha ha!

Pick a theorem, a program, whatever you want. My algorithm will tell you if it can be improved, and how!

No more need for consensus. Gone, the sluggishness and timidity of the scientific community! Can you even imagine?

That's what my algorithm is for!

Panel 1
Ninety-seven proofs for "P = NP" have been published. The scientific community didn't go after three-fourths of them.

Panel 2
All because the handful of specialists even able to understand such proofs were too lazy to look into it!

Someone might've solved it, and no one would know.

Panel 3
Absurd, isn't it?

Panel 4
My algorithm could, in a decisive and objective way, correct the work of anyone at all, known to the scientific community or not.

Ha ha ha ha!

Panel 5
What's so funny?

Panel 6
Well, that doesn't work out at all!

Panel 7
Hmph! What would you know about algorithmic complexity?

Panel 8
Nothing. Just common sense...

In order for your algorithm to be used, the scientific community has to greenlight it...

Panel 9
You need the community before you can *not* need it...

Hilarious, right?

I...	Do you really think a program is capable of creativity?
How do we know your algorithm works, anyway?	I'm sure of it. It works! It solves the traveling salesperson problem. P = NP!

OK, let's say P = NP. What does that change?

It means my algorithm can answer almost any question put to it in computing language.

I tested it...

It corrected and perfected the work of Han Fastolfe, Simon Burrick, Victor Blumberg, and even Vilhem!

What it did was sow chaos at the Institute!

Burrick and Blumberg are dead!

Fastolfe's robots were attacking him earlier.

And Vilhem's slated to die in ten minutes!

None of that's my fault! It's their research. All I did was improve it...

So what's with the gun? Where's Vilhem?

What, this? Vilhem gave it to me!
When he saw the prediction, he thought I was going to kill him. He came to threaten me...
I managed to reason with him, and he left for the boiler room.
He gave me the gun because he was afraid of doing something stupid...
Are you fucking kidding?!
His program only works within the confines of the Institute. He can evade the prediction by escaping.
You should stop chasing him. You're endangering him.
Nobody move!
It's over. Come with us!

"Shit, he's got a gun!" / "Paniandy. Drop your weapon!"	"What? Who are you?"	"Clean-up crew. Drop the gun or we'll have to open fire!"
"Did you call them?"	"You forced me to! If you hadn't killed two people..."	"Put the gun on the ground right now!"
"It was the algorithm. I didn't..." / **BANG!**		
"Fuckin' A..." / "He'd've shot!"	"Paniandy's dead!"	"She's gone. Gotta round her up. No bloodbath, OK?"

Why'd I call them? Those guys are insane!
They're totally going to kill Vilhem. I have to find him!
CLIC
Vilhem?
Give up! It's over!

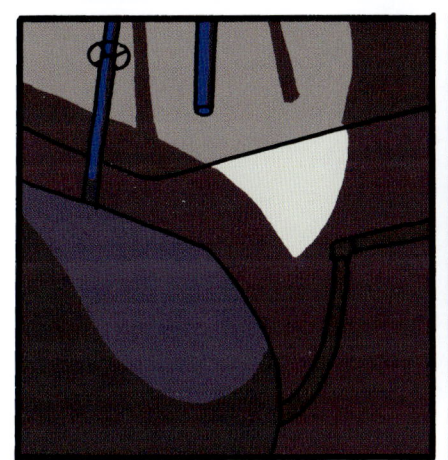

Dead end.

There must be a way...

Klak!

Vilhem?

— That's enough! Come with us!

— Grab her!

— You're hurting me!

— Mission accomplished!

— You gotta find Vilhem! He's in danger!

— We'll take care of him.

— What's that mean?

— It means Institute no. 4 is over! You're all going home!

— Stéphane! You see Vilhem?

— Quiet!

BAM! BAM

— Move out. The new team's coming!

One month later...

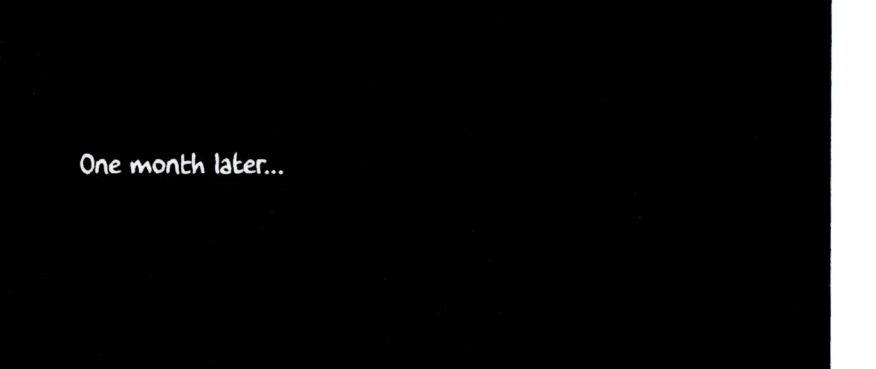

University Research Lab

There... Room 108.

Great. Could I have the wifi password?

In the file!

Ok, thanks!

This is a change...

You have 1 new message.

FROM: f.noel@lemonde.fr

Madame,
Your "Institute for the Study of Complex Systems" does not appear in any records. Your ravings about a program that predicts the future and a murderous algorithm are utter nonsense. The scientific community has already disputed Paniandy's proof. Vilhem Van de Vocht is no doubt simply avoiding you. I'd do the same if I were him.

Please stop harassing me!

François Denoël
Reporter, *Le Monde*

Panel 1: What? Bullshit!

Panel 2:
P VS NP
Vinaiy Paniandy's blog

Sign in

Username: Vinaiypan
Password: *******

Panel 3:
Proof #15 -(0 message)
Proof #16 -(0 message)
Proof #17 -(53 messages)
"What if P equaled NP?" -(1223 messages)

Panel 4:
T. Blanchard wrote:
There's no reason this algorithm should work in light of the flaws in its proof!

P. Richardson wrote:
One thing's for sure, at any rate: the logic errors on p.12 and p69 make it unlikely to work.

Andrew.G wrote:
Paniandy clearly has no mastery of his subject!

Panel 5:
Paniandy clearly has no mastery of his subject!

Gödel78 wrote:
I find it quite strange that everyone is suddenly interested in this proof only to discredit it! Do the results frighten you? Whatever the case, no one has managed to find a flaw in the algorithm!

Panel 7: So you smoke now?

Panel 8: When I'm nervous. Thanks for coming.

Panel 9: You'll be fine. It's not your first lecture!

Panel 10: I'll never get used to it.

"In the end, it's kinda like the Institute, huh?"	"A bit less luxurious, but..."	"Well... here goes."
"Wait!"	"What?"	"Paniandy's algorithm. Everyone says it's riddled with errors... Weird, right?"
"Wondered when you'd bring that up."	"But it works, dammit! He improved the residents' research!" "Not mine..."	"It enabled Vilhem to predict his own death!"
"Will you quit saying he's dead? No one ever found a body." "Precisely!"	"You know him as well as I do. He's not the kind to send postcards."	"Meanwhile, the next Institute is headed straight for chaos! How long before someone dies?!"

Vilhem was supposed to die at 21:21:12 exactly. He can't have left the prediction-effective zone in time!
Louise, I'll break it down for you.
If Paniandy's algorithm works, Vilhem died that night. If it didn't, he's alive somewhere and doesn't care about you. Which do you prefer?
The truth!
Then go on, study algorithmic complexity theory. You might have an answer in 20 years.
Until someone notices a flaw in your logic in turn, and you start doubting again!
So, what?! Are you saying I should stop thinking about all this?
Mysteries are a dime a dozen. All I'm saying is pick one you have a shot at solving.
Quit freaking out and go back to linguistics!
I'm late...
See ya.

— Mademoiselle François!

— Why, if it isn't Monsieur Raynaud!

— How's the article coming?

— Er... slowly but surely!

— I'm certain you will be as fine a professor as you are a scientist!

— Well. My lecture hall awaits.

— Tell me all about your research over coffee soon!

— Oh, yeah, sure.

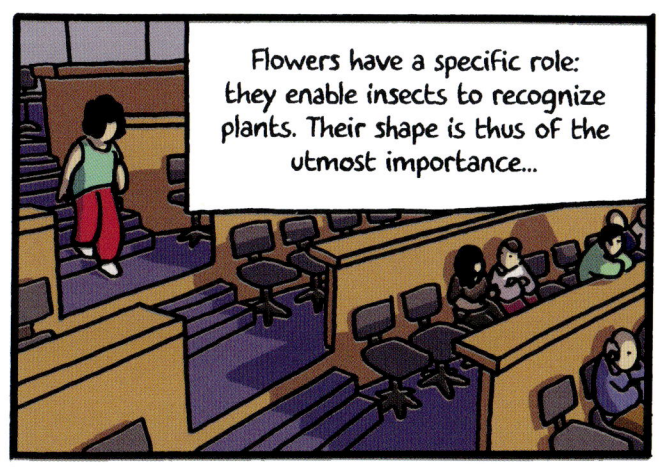

Flowers have a specific role: they enable insects to recognize plants. Their shape is thus of the utmost importance...

Conversely, insects do not use leaves to recognize plants. So the question is...

...why is there such a variety of leaf shapes?

Well, I have no idea...

What interests me here isn't "why" but "how."

Let's assume leaf shape is, above all, the result of purely mechanical limitations.

The leaves I've studied meet two criteria...

leaf

They're folded in the bud, taking up all the available space.

So we have a surface that's folded and limited by a given space.

klik!

By varying fold shape and bud size, nature succeeds in making very different leaves.

Panel 1: Interesting, how you can get the same results with scissors and a mere sheet of paper.

Panel 3: I have here two sheets folded at different angles.

Panel 4: A little snip...

Panel 5: And ta-da: a maple leaf and a fig leaf.

Panel 6: So much for my intro...

Panel 7: And now...

FOLDS AND FORMS OF LEAVES

Chapter 1

Bud Structure and Leaf Symmetries

Panel 8: Let's cut to the chase.

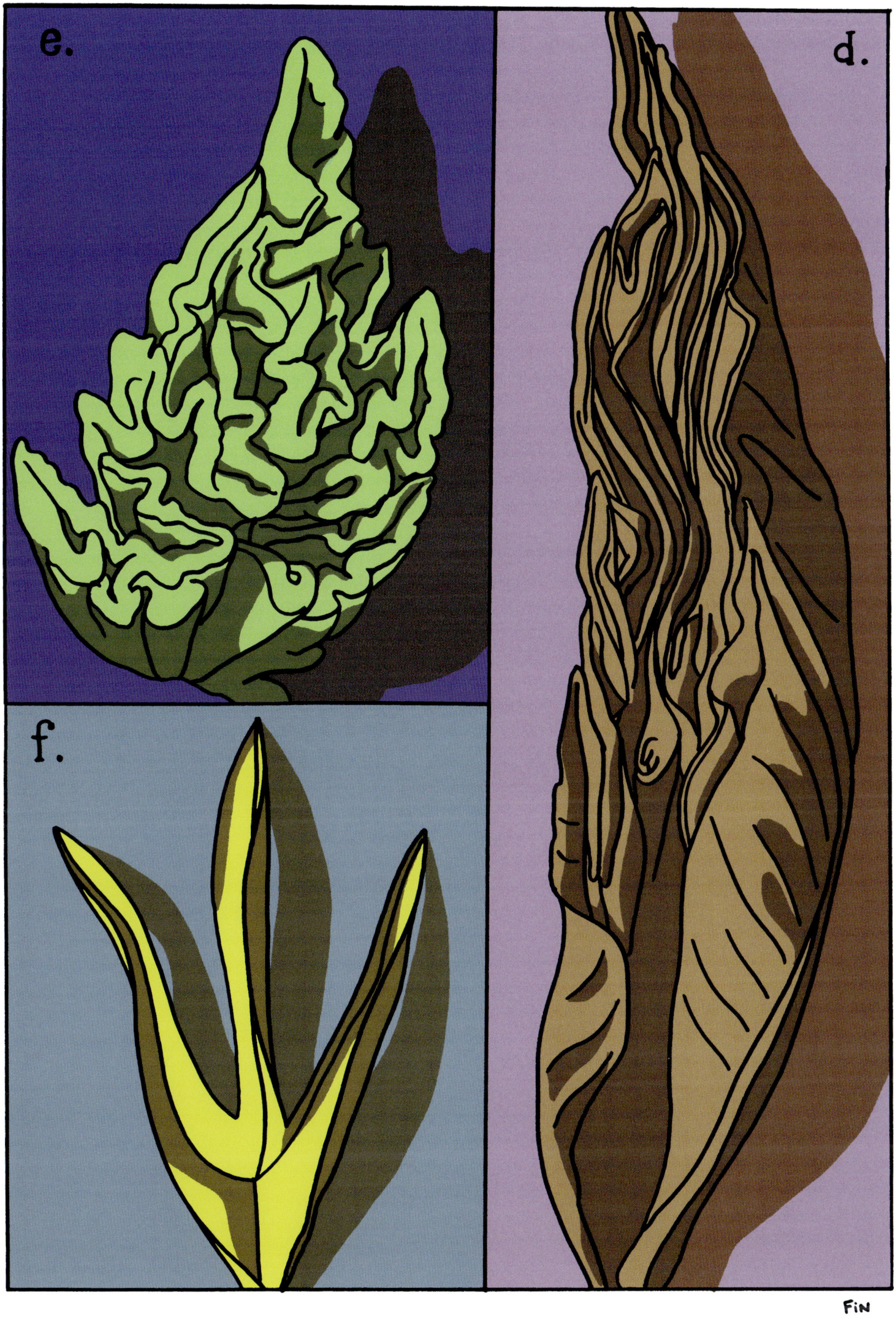
e.
d.
f.
FIN

Hats off to Stéphane Douady and Étienne Couturier
for the beauty of their research on natural shapes and forms.

© 2023 Robin Cousin and Éditions FLBLB

Originally published as Le Chercheur fantôme © 2019 Robin Cousin and Éditions FLBLB.

MIT Press edition published by arrangement with eddy agency.

All rights reserved. No part of this book may be reproduced in any form by any electronic or mechanical means (including photocopying, recording, or information storage and retrieval) without permission in writing from the publisher.

This book was set in ROBIN-Plume and Madras by the MIT Press. Printed and bound in the United States of America.

Library of Congress Cataloging-in-Publication Data

Names: Cousin, Robin, author, artist. | Gauvin, Edward, translator.
Title: The phantom scientist / Robin Cousin ; translated by Edward Gauvin.
Other titles: Chercheur fantôme. English
Description: Cambridge : The MIT Press, [2023] | Originally published as Le Chercheur fantôme © 2019 Robin Cousin & Éditions FLBLB.
Identifiers: LCCN 2022018576 | ISBN 9780262047869 (hardcover)
Subjects: LCSH: Scientists—Comic books, strips, etc. | System theory—Comic books, strips, etc. | LCGFT: Thriller comics. | Graphic novels.
Classification: LCC PN6747.C67 C4713 2023 | DDC 741.5/944—dc23/eng/20220429
LC record available at https://lccn.loc.gov/2022018576

10 9 8 7 6 5 4 3 2 1